SOLAR CELLS

ABOUT THE AUTHOR

Jeffrey A. Mazer, Ph.D., is a semiconductor device engineer in suburban Maryland.

Cover Photo: Photovoltaic array supplying power for a microwave repeater at Iron Mountain, Idaho. *(Photo courtesy Applied Power, Lacey, WA)*

Jeffrey A. Mazer

SOLAR CELLS:
An Introduction to
Crystalline Photovoltaic Technology

KLUWER ACADEMIC PUBLISHERS
Boston / Dordrecht / London

Distributors for North America:
Kluwer Academic Publishers
101 Philip Drive
Assinippi Park
Norwell, Massachusetts 02061 USA

Distributors for all other countries:
Kluwer Academic Publishers Group
Distribution Centre
Post Office Box 322
3300 AH Dordrecht, THE NETHERLANDS
ISBN-13:978-1-4613-8066-5   e-ISBN-13:978-1-4613-0475-3
DOI:10.1007/978-1-4613-0475-3

Library of Congress Cataloging-in-Publication Data

A C.I.P. Catalogue record for this book is available
from the Library of Congress.

CONTENTS

LIST OF SYMBOLS

α .... absorption coefficient

η .... conversion efficiency; impedance

λ .... wavelength

μ_n, μ_p ..... electron and hole mobilities

ρ .... resistivity

ρ_s .... sheet resistance

τ .... minority carrier lifetime; dielectric relaxation time

ϕ .... electrostatic potential

A .... area

c .... speed of light

C_0 .... solid solubility concentration

C_s .... surface concentration

D .... diffusivity; electric flux density

E_C, E_V .... conduction and valence band edges

E_F .... Fermi energy

E_{FN}, E_{FP} .. quasi-Fermi energies for electrons and holes

E_g .... bandgap

E_i .... Fermi energy in intrinsic material

FF .... fill factor

h .... Planck's constant

I .... current

I_D .... dark diode current

I_0 .... dark diode saturation current (pre-exponential factor)

I_{ph} .... photogenerated current

I_{QNR} .... QNR component of dark current

I_{QNRO} ...... QNR saturation current (pre-exponential factor)

I_{SC} .... short-circuit current

I_{SCR} .... SCR component of dark current

I_{SCRO} .... SCR saturation current (pre-exponential factor)

J .... current density

J_0 .... dark diode saturation current density

J_{SC} .... short-circuit current density

k .... Boltzmann's constant

L_n, L_p .... minority electron and hole diffusion lengths

N_{AA}, N_{DD} ... acceptor & donor concentrations

N_C, N_V ..... densities of states in conduction and valence bands

n_i .... intrinsic concentration

n_0, p_0 .... electron and hole concentrations in thermal equilibrium

n_p .... electron concentration on the p-side

n, p .... electron and hole concentrations

n_{po} .... electron concentration on p-side in thermal equilibrium

p_n .... hole concentration on the n-side

p_{no} .... hole concentration on n-side in thermal equilibrium

q .... electron charge

R_s, R_{sh} ...... series and shunt resistances

S .... surface recombination velocity; spectral distribution

V .... voltage

V_{bi} .... built-in voltage

V_N, V_P .... quasi-Fermi potentials for electrons and holes

V_{OC} .... open-circuit voltage

$\Delta n, \Delta p$ ..... excess electron and hole concentrations

LIST OF FIGURES

PREFACE

The expense of extending the conventional electric power utility-grid to remote locations (about $20,000 per mile in the U.S.) often prohibits the installation of electric lighting, common household appliances, television receivers, and other telecommunications equipment in such locations. Additionally, the installation of important, but perhaps seldom noticed, electronic equipment such as remote automatic weather monitoring stations, microwave telephone repeaters on mountain tops, and earth-bound navigational aids for commercial aircraft and ships, is also impeded by the difficulty in providing electric power. The unavailability or expense of electric power in remote locations is a particularly acute problem in underdeveloped countries. The opportunity for people in these countries to improve their lives through technology will be mostly lost if they can not obtain even modest amounts of electricity. In a rural village, for example, just one or two kilowatts of electrical power can make a tremendous difference in the quality of life by providing refrigeration for food and medicinal storage, lighting for reading after dark, television reception, and water pumping and purification. In the industrialized countries, there is the additional problem of the environmental burden of producing and distributing huge amounts of electricity in an economy which is always hungry for electric power. However, within the last several years, photovoltaic (i.e., *crystalline silicon solar cell*) engineering has become a cost-competitive approach for ameliorating certain difficult electrical power needs in both underdeveloped and industrialized countries.

Aside from independent advances in solar cell device theory (mostly at universities), silicon solar cell technology has been the beneficiary of numerous advances in semiconductor materials engineering and processing technology. This has been fostered by the closely related and very mature microelectronics industry, i.e., the silicon integrated circuits industry. As a result, crystalline silicon cells (both single crystal and polycrystalline) have been under continuous development since 1954 when the first practical solar cell was produced at Bell Laboratories. Crystalline silicon solar cell technology is now undergoing rapid commercial expansion as low-cost materials and semicontinuous manufacturing methods become well established. This is evidenced by the strong growth in worldwide shipments of crystalline silicon photovoltaic modules over the last three years: from 41 megawatts in 1991 to 81 megawatts in 1995. While non-crystalline silicon photovoltaic technologies are being aggressively developed (and, in the case of amorphous silicon, are already in the commercialization stage), the advantages and manufacturing maturity of crystalline silicon cells almost guarantee that this segment of the industry will dominate the photovoltaics market at least through the year 2001.

This book, written for an engineering audience, provides an introduction to the theory and fabrication of crystalline silicon solar cells and modules. It is intended to fill some of the existing gap between rigorous treatments of solar cells as semiconductor devices and the numerous simpler descriptions of solar cells that present little, if any, discussion of physical device mechanisms. The emphasis here is placed on an explanation of physical concepts, rather than on mathematical rigor. There are relatively few equations, and a prior understanding of semiconductor devices is not required. Engineers, technicians, and project managers who are new to the solar cell field will hopefully gain an understanding of the crystalline silicon cell technology. Marketers and sales personnel will find selective parts useful.

There are five chapters oriented around numerous figures. The first chapter introduces the reader to photovoltaic terminology and discusses some of the salient issues of solar cell development and commercialization. Solar cells are first and foremost semiconductor devices. Therefore, the second chapter explains basic semiconductor principles, including the fundamental structural element of all commercial silicon solar cells, namely, the pn-junction. With the first two chapters as a background, the third chapter explains how solar cells function, provides the equivalent-circuit model analysis of their operation, and describes the numerical "figures of merit" that characterize solar cell performance. Chapter Four discusses cell and module development. Manufacturing technology is quickly changing as production levels increase. Thus, the discussion does not dwell on the details of specific processes, but rather, on the general approach to solving the generic problems encountered in making cost-effective cells and modules. The fifth chapter discusses non-ingot silicon and novel technologies. Because of the cost of sawing ingots and the attendant material waste, non-ingot technologies probably hold the best commercial prospects for the long term. These include crystalline silicon ribbons and thick films deposited on inexpensive substrates, both of which have experienced rapid development in the last several years. Gallium arsenide and indium phosphide cells are also discussed. These are crystalline technologies displaying both high laboratory efficiencies and enhanced radiation hardness for space applications. Gallium arsenide also has great promise for terrestrial concentrator systems. Last, there is a discussion of thermophotovoltaic (TPV) devices. These photovoltaic cells incorporate materials and designs to make use of an artificial infrared spectrum. While they could be used for solar photovoltaics, their development is intended for optical coupling to a narrow-band thermal source, for example, a very hot ceramic emitter heated by burning natural gas. TPV presents exciting possibilities for stable and reliable remote power supplies that transform fossil fuel directly into electric current.

The author gratefully acknowledges Dr. Bolko von Roedern for his many helpful comments about the manuscript, and Ms. Lani Johnson for her patient and excellent rendition of the drawings and camera-ready format. Any mistakes in the text or drawings are strictly the author's.

1

AN OVERVIEW OF SOLAR CELL TECHNOLOGY

1. INTRODUCTION

Solar cells, sometimes referred to by the more general term *photovoltaic (or PV)* cells, are semiconductor devices that convert sunlight directly into electric current to produce useable electric power. These devices were first fabricated and extensively studied in the mid 1950s[1]. They were made from crystalline silicon wafers. Initial efficiencies were only about 6%, and early solar cells were treated mostly as laboratory curiousities. The first significant application was in the form of spacecraft power supplies with the rapid development of the U.S. space program in the late 1950s. Since then, knowledge of silicon device physics and fabrication methods has greatly expanded the future of photovoltaics. Today, crystalline silicon remains the dominant photovoltaic material. It is cells made from this and other crystalline semiconductors that form the topic of this book.

An immense amount of solar energy is radiated toward the earth. The power density (i.e., power per unit area) of sunlight is approximately *135 mW/cm²* just outside the earth's atmosphere. This fact has led to some ambitious ideas for supplementing conventionally generated electric power with photovoltaic generated power. For example, in the early 1970s, the Arthur D. Little Company developed a plan for the assembly (in outer space) of a huge solar cell array. The array would then be launched into geostationary orbit about 36,000 km above the earth. The current generated by such an array would be converted into microwave radiation and transmitted to earth for collection by a large-area ground antenna, with subsequent conversion back into useable current. With a 10% total system conversion efficiency, a 100 km² array would yield about 13.5 gigawatts of peak power. This represents about 4.2% of the time-average U.S. electric power consumption (about 320 gigawatts) in the year 1993. While this formidable engineering task was never attempted, its conception illustrates the potential for tapping a vast amount of power through photovoltaics.

Over the years, solar cell designs have become specialized for either terrestrial or outer space applications. Except for certain very low power applications such as watches and cameras, terrestrial solar cells always appear in the form of a *module*. A module is a group of solar cells that are wired together in a suitable electrical configuration and then hermetically packaged in a weather-proof flat container. One side of the module is transparent, allowing sunlight to reach the solar cells. The module has electrical leads for connection to a load, and represents the minimal deployable arrangement of solar cells for a practical terrestrial application. Such applications include a stand-alone module or a small array of modules to power isolated systems. Common applications are water pumping and home electrification in remote villages, microwave telephone repeaters

on mountain tops, remote weather recording and data transmission stations, emergency telephones, and street lights (fig. 1.1-1). For outer space applications, hermiticity is not required. However, the cells must be fabricated to withstand the hostile high-energy nuclear particle and ultraviolet radiation environment found outside the atmosphere. Almost all U.S. spacecraft have used solar cell arrays for their long-term electrical power needs.

Figure 1.1-1(a) Microwave repeater at Clay Hills, Utah. (*Photo courtesy Applied Power, Lacey, WA*)

Figure 1.1-1(b)   Pumping water for livestock at Dove Creek, Colorado, using a tracking array with two MSX-60 modules. (*Photo courtesy Solarex*)

Figure 1.1-1(c)
A 7-kW array
provides power
to a home in
Carlisle, MA.
(*Photo courtesy
Solarex*)

Figure 1.1-1(d)   Emergency roadside telephone in Jakarta, Indonesia.
(Photo courtesy Solarex)

Figure 1.1-1(e)   The 3-kW Solartrogen Home using 50 Solarex MSX-60 modules mounted on garage roof. (*Photo courtesy Solar Depot, San Rafael, CA*)

Figure 1.1-1(f)   300-kW array being installed on a building at Georgetown University, Washington, DC. (*Photo courtesy Solarex*)

Figure 1.1-1(g)   A large photovoltaic array supplies power to a
communications facility. (*Photo courtesy SEIA*)

Figure 1.1-1(h)   Utility-connected 200-kW photovoltaic array at Davis, CA. (*Photo courtesy Siemens Solar Industries*)

Modules and small arrays are most often used to charge batteries, rather than directly supply power for the end-point application. The reason for this is twofold.

(i) With final applications like low-power lighting, remote electronic instrumentation, or water pumping equipment, a steady-state direct current is required.

(ii) Since most of these applications must also be functional at night, batteries must be interfaced with the photovoltaic source to assure continuous power availability. Solar cells are inherently well-suited to low-voltage dc applications. They can be used in almost any terrestrial system where a battery needs to be periodically recharged.

The alternative to charging batteries is the direct interfacing of an array to the commercial electric utility grid. In this configuration, the direct current of the array is converted to an alternating current through an inverter, and synchronized to the 3-phase voltage of commercial power lines. The array feeds electricity into the power grid.

Energy that is fed back into the system can be recorded by a meter so that the owner of the array receives a financial credit from the local power company. Private users are not the only organizations that have considered using PV in this mode. As of the mid 1990s, a number of utility companies have started prototype large-scale PV arrays for *central station* generation of electric power. Several arrays of hundreds of kilowatts or larger have already been deployed (fig. 1.1-2).

Figure 1.1-2(a)   1-MW utility validation array at Davis, CA. (*Photo courtesy SEIA*)

Most semiconductor environmental issues relating to silicon solar cell processing have already been solved by the integrated circuits industry. As in the processing of other semiconductor devices, caution must be used in disposing of various toxic aqueous solutions and diffusion gases that are used during fabrication. But this is a minor problem and does not represent a major difficulty in the development of a large-scale industry. The passive operating qualities of all solar cells (i.e., noiseless, pollutionless, non-radioactive, very low maintenance), have stimulated hopes that these devices might someday become economically feasible for large-scale central station power generation. This was particularly the case after the 1973 and 1979 oil embargoes. Unfortunately, when compared to fossil fuels, the relatively high costs of semiconductor materials and processing have been a serious impediment to such implementation. Most photovoltaic power generation deployed as of 1996 is still in the form of small *distributed* systems (typically less than 1 kW) for remote applications.

The common figure-of-merit for evaluating the costs of complete, installed, photovoltaic systems is dollars per kilowatt hour ($/kWh). However, in talking about module costs, one usually quotes the cost per peak watt ($/W$_p$), where the term "peak watt" refers to module output at maximum test illumination – typically 100 mW/cm^2. This means that the figure-of-merit for a complete installed system is dollars per unit energy, whereas, the figure-of-merit for a module is dollars per unit power.

As a comparison of system costs between conventional centralized generation and distributed photovoltaic generation, the average residential price of utility-generated electric energy in the U.S. in 1994[2] was *$0.079 per kWh* (for a yearly total of 2.9 x 10^{12}

Figure 1.1-2(b)   2-MW photovoltaic installation at the Rancho Seko Nuclear Plant, CA.
(Source unknown)

kWh). This is opposed to between *$0.20 and $0.40 per kWh* for the very best case of high-insolation PV amortized over a twenty to thirty year period. High-insolation (i.e., sunny) locations receive more total solar energy per year than, say, Helsinki, Finland. The insolation in northern Europe is approximately half that of the California desert, and thus, the amortized cost per kilowatt-hour is at least twice as much. Module manu-facturing cost and the subsequent cost per peak watt, on the other hand, are not depen-dent on insolation. In 1995, module manufacturing cost and large-volume vendor price were roughly $5/W$_p$ and $4/W$_p$, respectively. Manufacturing cost includes deprecia-tion of equipment, return on investment, and R&D expenses. However, the immediate cost of material and labor was approaching $2/W$_p$. The total vendor price for complete installed systems was roughly $8 to $12 per peak watt, depending on the configuration (off-grid or grid-connected), the technology, and the volume.

The spread of values for PV system energy ($/kWh) is large because of the variable cost for the necessary BOS (balance of systems) equipment such as storage batteries, charge controllers, and inverters which might be required for a specific photovoltaic application, and because of the geographical dependence of the average annual illumi-nation (insolation). However, this critical figure-of-merit ($/kWh) is steadily improv-ing. Module and BOS manufacturing costs are steadily decreasing. And, like many other industries, photovoltaics can benefit from economy of scale. One producer of concentrator modules estimates that a 300-MW array would bring the overall system energy price down to between 4 and 7 cents per kWh. Such large-scale arrays have not yet been built, but it is clear that by the year 2005, improvements in both module and BOS manufacturing efficiencies might very well make energy from large PV arrays cost competitive with conventional central station generation in certain parts of the U.S. where the insolation is high.

Historically, the great majority of commercial solar cells (and microelectronic circuits, for that matter), have been made from *silicon* (Si), a cheap and plentiful element found in sand and other minerals. Aside from the abundance and several outstanding elec-tronic and chemical properties of silicon, silicon solar cells have been the beneficiary of the highly mature silicon integrated circuit processing technology that has been un-der continuous development since the 1950s. Four material properties, in particular, make silicon a desirable choice for solar cell manufacture. These properties relate to the chemical insulation and surface passivation qualities of its common oxide, silicon dioxide (SiO_2), and the electronic properties of pure silicon material.

Silicon dioxide is an excellent electrical and chemical insulator, even as a thin film less than a micron (1 μm) thick. The ability to form a thin film of silicon dioxide provides the electrically insulating layer necessary in certain high efficiency silicon cell designs. As a chemical insulator, silicon dioxide resists organic solvents that are used in photo-lithographic processes during fabrication of high-efficiency cells. The oxide is water insoluble (unlike the oxide of germanium) and is etched by few mineral acids other than solutions of hydrogen fluoride (HF). While the insulation properties of silicon dioxide are absolutely essential in the fabrication and performance of silicon integrated circuits, an even more important property from the solar cell standpoint is the ability of the oxide to electrically passivate the surface of a cell. Passivation relates to the ability of a thin layer (0.15 nm) of silicon dioxide to surpress the parasitic current associated with the front surface of the cell. The oxide "closes" the surface in the sense that it ties up the unused or "dangling" silicon bonds appearing at the abrupt termination of the silicon crystal at the wafer surface. This fact alone has a significant positive effect on cell efficiency. Additionally, the hard oxide layer protects the finished cell from con-tamination and surface damage. While silicon readily forms a native oxide several tens of angstroms thick, a clean film up to a micron thick can be easily formed either by *thermal oxidation* of the silicon wafer in an oxygen or steam ambient, or by *chemical vapor deposition* (CVD) techniques.

The second remarkable property of silicon is the *high electrical resistivity* of the pure material. This implies a low crystal defect and metal impurity density can be achieved in a silicon crystal. It is important because it obviates the need for the expensive intro-duction of intentional contaminants (dopants) to achieve high resistivity. Other semi-conductors, e.g., gallium arsenide, are not always as flexible in this regard. High resis-tivity starting material allows a high shunt resistance (1 kilo-ohm or more) that pre-vents shorting between the front and back contacts. This facilitates many different cell designs.

A third property is the relatively *low saturation current density* that can be realized in a silicon cell. Saturation current (discussed in Chapter 3) is both a material- and design-dependent parameter. While inherent in any solar cell (made from any semiconductor material), the saturation current is strictly parasitic and must be minimized to achieve reasonably efficient operation.

Finally, the fourth property is the ability of crystalline silicon to absorb a fairly large part of the solar spectrum, given thick enough material. Absorption is dependent on the type and magnitude of a material parameter called the *bandgap* or *energy gap*, E_g. The bandgap for silicon is about 1.1 eV at room temperature (300 K) and allows the silicon crystal, at least in principle, to absorb solar radiation well into the infrared part of the spectrum. This is to be contrasted with gallium arsenide which has a bandgap of about

1.4 eV, and can not absorb in the infrared – a fact which is particularly significant for earth-bound cells for which the sunlight is relatively rich in infrared light. The above four properties make silicon an excellent material for photovoltaics.

By way of comparison, *gallium arsenide* (GaAs), affords the advantages of higher efficiency and higher operating temperature (both attributable to a large and "direct" bandgap), and greater radiation resistance (radiation hardness) than silicon. While GaAs is *much* more expensive than silicon, and harder to work with due to its brittleness, the maturity of the GaAs processing technology during the late 1980s has allowed this material to become a viable option for space cells. Deposition of thin films of GaAs on inexpensive substrates could also have terrestrial application. This is particularly plausible in concentrator systems employing illumination intensities of several hundred suns. Specially designed cells employing thin films of indium phosphide (InP) are also being developed for use in high-radiation space environments because of the extreme radiation hardness of this material. It should be noted that a small percentage of solar cells are also made from alloys of copper indium diselenide ($CuInSe_2$) and from cadmium telluride (CdTe) material systems. These last two materials show promise for high throughput semi-continuous processing, and they may become major photovoltaic options in the early part of the next century. Overall, though, silicon is still by far the most common material for solar cells. *In 1995, at least 98% of photovoltaic module production was done in silicon, and about 85% was in crystalline silicon.*

Yearly production of PV modules is steadily increasing. In 1995, worldwide module production totalled about *81 megawatts* (MW)[3], up from about 29 MW in 1987. From this trend, industry analysts are projecting about 500 MW total installed photovoltaic generating capacity in the U.S. by the year 2000. This is equal to the power output of a medium-sized fossil fuel or nuclear power plant.

Regardless of its future for large-scale centralized power generation, the most significant terrestrial application for PV is probably *small-scale power generation for villages in underdeveloped countries.* By periodically recharging a bank of storage batteries, an array of relatively low-cost PV modules can supply power for refrigeration, lighting, health clinics, television receivers, water pumps, and food processing equipment that immeasurably improve the quality and productiveness of people's lives.

The most common approach to making solar cells has been the diffusion of a thin layer (~ 0.4 to 0.6 microns) of suitable impurity into a wafer of silicon to form a large-area *pn-junction diode.* A pn-junction diode is a two-terminal structure formed by the close contact of a free-electron-rich layer of semiconductor with a free-electron-deficient layer of semiconductor. While there are other generic solar cell structures, the pn-junction diode is by far the most common generic form. Consequently, an understanding of solar cells requires an understanding of the properties of semiconductor materials as well as some of the device physics of pn-junction diodes. These concepts are explained in Chapters 2 and 3. Chapter 4 looks at approaches for fabricating solar cells and photovoltaic modules. Chapter 5 looks at some novel technologies and discusses the very interesting emerging technology of thermophotovoltaic (or infrared) cells.

1.2 EXPLOITATION DEPENDS ON THE DEVELOPMENT OF CHEAP MATERIALS

The problem of relatively poor performance-to-cost ratio is fundamentally related to the fact that solar cells are *large-area devices* (fig. 1.2-1). Medium-efficiency solar cells are typically between 100 and 225 square centimeters in area. This is in contrast to integrated circuits (ICs) which often contain a microprocessor or other sophisticated system in an area of about 1 cm². Because of the relatively low power density of sun-

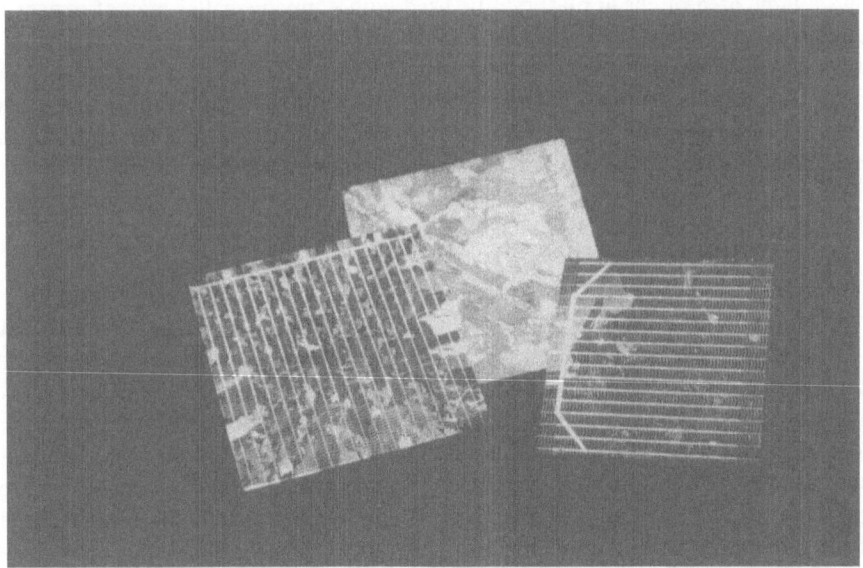

Figure 1.2-1   Solar cells, unlike their microelectronic circuit cousins, are large-area devices.

light (i.e., power/cm^2), many individual cells are required for most applications. The individual cells are arranged into modules which are then wired in an appropriate parallel and series combination to produce the required current and voltage output for a given load. A medium-efficiency (12%) photovoltaic module capable of producing 120 watts of peak power (approximately the power requirement of two household incandescent light bulbs or three fluorescent bulbs) would require an area of about 1 m^2.

The salient point is that most solar cell applications require *a lot of material*. In IC manufacture, the cost driver is the complexity of the process, i.e., the number of photolithography steps, the cost of the circuit packaging, and (sometimes) the time required for circuit testing. None of these is strongly dependent on area. In contrast, the major cost drivers for photovoltaics are the amount of semiconductor material for the cells and the material for the construction of the module, both of which are strongly dependent on area.

This fundamental aspect of photovoltaic producer cost can be partly relieved with the use of inexpensive grooved plastic flat lenses (*Fresnel lenses*) to create *concentrator systems* that focus sunlight onto a small area of semiconductor. Such an approach allows a large area of cheap plastic to be used with a much smaller area of expensive semiconductor material. The trade-off is that concentrator systems require close one- or two-dimensional tracking of the sun in order to be effective. Additionally, concentrator systems are often unacceptable because they lack the long-term maintenance-free reliability required by many photovoltaic applications in remote locations. In any case, concentrator systems are only feasible in those geographical locations that have infrequent cloud cover, i.e., direct sunlight during most of the day. The southwestern U.S. is a good geographical candidate for such systems (fig. 1.2-2).

There are two alternative approaches for relieving the problem of material costs. One is to use *thin films* of high-absorption coefficient semiconductor such as cadmium telluride, copper indium diselenide alloys, amorphous silicon, or gallium arsenide and its alloys. The other is to use a *thick film* of crystalline silicon deposited on an inexpensive substrate and to incorporate light trapping techniques that will cause multiple reflections within the silicon layer. The thin- or thick-film approach has the inherent advantage of requiring much less semiconductor material than the conventional bulk silicon approach. As a measure of comparison, in the photovoltaic terminology, thin films are about 1 to 5 μm thick, thick films are about 30 to 100 μm thick, and bulk silicon material (either ribbon or ingot-based) is about 250 μm thick. Bulk Si cells used in space must mnimize mass and are thinner – about 100 μm. However, semiconductor cost is only one of several major cost drivers for photovoltaics. None of the above

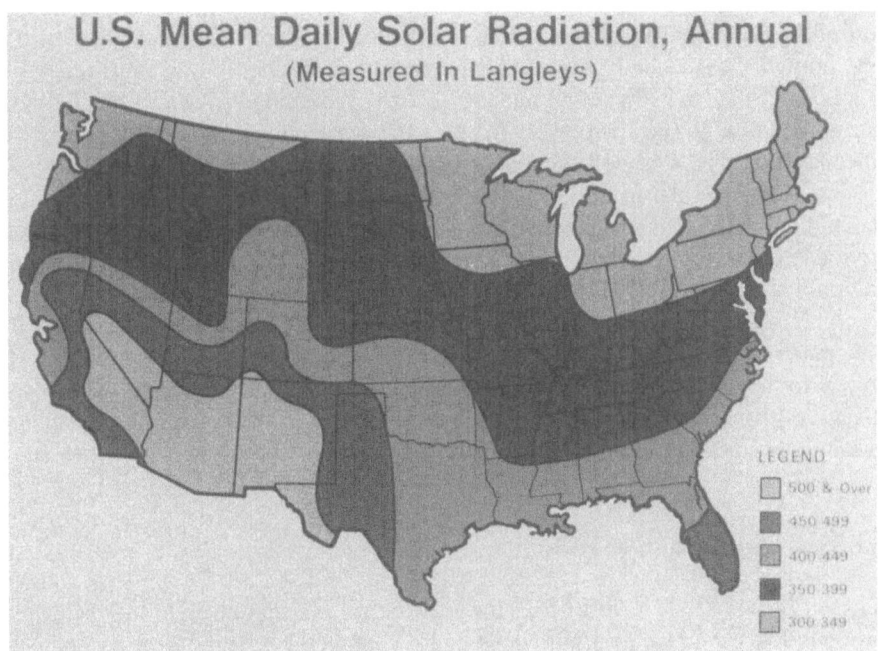

Figure 1.2-2   U.S. annual mean daily solar radiation measured in langleys.
1 langley = 1 cal/cm² = 4.19 J/cm². (*Photo courtesy Solarex*)

alternative approaches, i.e., concentrators with Si or GaAs, thin non-silicon films, or thick films of Si with light trapping, is sufficient by itself in drastically reducing photovoltaic producer prices.

An important performance parameter in the characterization of solar cells is *power efficiency*, i.e., the ratio of electrical power output to total solar power input. Contrary to popular belief, it is usually not the efficiency of solar cells which dictates cost feasibility, but rather the costs of the materials and the processing of the materials into cells and modules.

There are three exceptions to this prioritization of material cost and efficiency. One is the rare case of high concentration terrestrial systems (100 X or greater) where the area of semiconductor material is so small that efficiency becomes the dominant performance-to-cost parameter. A second exception is the case of solar cells for spacecraft. For spacecraft, the cost of the solar array is usually very small compared to the overall cost of the mission. The mass and area of cells that can be launched and deployed is very limited. Consequently, the priority for space cells (aside from extreme reliability in a hostile solar-radiation environment) is high conversion efficiency coupled with low mass, I.e., high specific power (W/kg). A third exception is the case of amorphous silicon cells. There is general agreement that thin-film technologies will eventually cut the cost of PV modules in half. However, production line amorphous silicon cells have stable illuminated efficiencies of only about 6% to 8%. As of 1995, the world record for stable efficiency for a laboratory grade amorphous silicon mini module (900 cm^2) was only about 10.2%. Clearer understanding of the device and processing physics of amorphous silicon cells may well boost stable production line cell efficiencies to the 10% range by the year 2001. This would greatly increase the annual sales of this technology for large-scale centralized power generation. Thus, because of the built-in cost advantage that comes with thin films, the development of the amorphous silicon cell industry appears to be strongly dependent on cell efficiency.

1.3 THE SOLAR SPECTRUM

As seen in fig. 1.3-1, most of the power in the solar spectrum is in the bandwidth between *0.30 μm* in the near ultraviolet and *2.0 μm* in the near infrared. By comparison, the visible spectrum extends from a wavelength of 0.4 μm to 0.7 μm. At the outer edge of the earth's atmosphere, the power density is 135.3 mW/cm^2. However, because ozone, water vapor, carbon dioxide, and other components of the atmosphere absorb strongly at various wavelengths in the spectrum, both the intensity and spectral distribution of sunlight are significantly changed as light passes through the atmosphere. Additionally, the spectrum seen by a solar cell depends on whether the light is direct or global. Direct light refers only to the orthogonal component of sunlight reaching the front surface of the cell. Global refers to light reaching the cell surface from all angles. At the earth's surface at sea level, with the sun at the zenith, the power density is attenuated to 92.5 mW/cm^2.

As light passes through the atmosphere, an increasingly larger fraction of the power is in the red end of the spectrum. Absorption of light is wavelength dependent, with the atmosphere absorbing blue light more strongly than red light. The spectral distribution of sunlight outside the atmosphere is referred to as *Air Mass Zero (AM0)*, because light passes through no air at all. At sea level, with the sun at the zenith, the spectral distribution is referred to as AM1 because the light has traversed a path length of one atmosphere thickness. When the sun is about 42° above the horizon, the path length is 1.5 atmospheres. The corresponding spectral distribution is referred to as AM1.5. The power

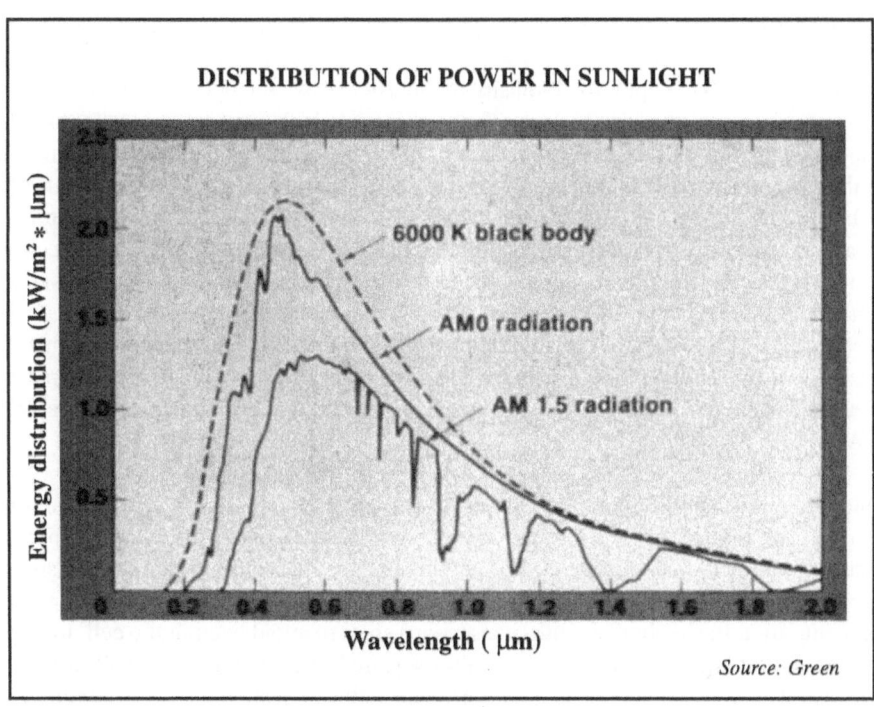

DISTRIBUTION OF POWER IN SUNLIGHT

Energy distribution (kW/m² ∗ μm)

6000 K black body

AM0 radiation

AM 1.5 radiation

Wavelength (μm)

Source: Green

Figure 1.3-1   Most of the power in the solar spectrum is between the wavelengths 0.30 microns and 2.0 microns. (*Photo courtesy SEIA*)

density at sea level when the sun is at this inclination is about 84.4 mW/cm². In general, AMx refers to the spectral distribution for a path length of x atmospheres (in the absence of cloud cover).

Solar cell performance *changes with spectral distribution* because absorption of photons by the semiconductor is wavelength dependent. The energy of a photon is given by $E = h\nu = hc/\lambda$, where h is Planck's Constant (6.63×10^{-34} J-s), ν is frequency, and λ is wavelength. When light reaches the interface between air and semiconductor, part of the light is reflected back into the air and part is transmitted into the semiconductor. The reflection and transmission of photons at the interface is a statistical process dependent on wavelength. Of the photons which are transmitted into the semiconductor, those that are sufficiently energetic – i.e., have energy greater than the bandgap, a parameter discussed in Chapter 2 – are absorbed by the semiconductor with a characteristic exponential decay length or skin depth. The skin depth is equal to $1/\alpha$, where a = $\alpha(\lambda)$ is the absorption coefficient. The photon flux (photons/cm²-sec) at a depth x from the surface is given by $P(x) = P_o e^{-ax}$, where P_o is the transmitted flux at the surface. At one skin depth, the intensity of a particular wavelength has fallen to $1/e$ or 37% of the intensity at the surface. Those photons which are less energetic than the bandgap pass through the material. For crystalline silicon, the skin depth at the blue end of the spectrum is much shorter than at the red end of the spectrum[4]($1/\alpha = 0.0096$ μm at $\lambda = 0.35$ μm, versus 5×10^6 cm at $\lambda = 1.4$ μm). Silicon very strongly absorbs in the ultraviolet and is essentially transparent in the infrared. Even at the visible limits of the spectrun, there is a great divergence of skin depth values: $1/\alpha = 0.10$ μm at $\lambda = 0.4$ μm versus 5.3 μm at $\lambda = 0.7$ μm. Thus, the characterization of solar cell performance must include the specification of the spectral distribution under which the cell is measured – not just the intensity of the sunlight. Solar cells designed for outer space applications are typically measured under AM0 laboratory conditions, while a common reference for terrestrial cells is AM1.5.

Output measurements are most often performed in a laboratory with an artificial light source. The light is passed through a series of filters to obtain the desired spectral distribution for the measurement. For any given spectral distribution, the intensity of the light source (power density) can also be adjusted. Measurements are made with a short-duration flash (about a tenth of a second) so as to avoid heating the cell. Regardless of the air mass, solar cells are often measured at an intensity of 100 mW/cm², because it is convenient for calculations. The industry test standard (*Standard Test Conditions*) that is referred to in product literature is 1000 W/m² irradiance, 25°C, and AM1.5 spectrum. Exact spectrum determination is defined through ASTM specification E891-87[5] and related specifications for spectral irradiance. This approach replaces earlier methods that were dependent on calibrated reference cells[6]. Standard Test Conditions are valuble as a reference that can be easily reproduced in a laboratory. In practice, though, when a module is deployed, the ambient conditions can be quite different from STC. Product literature usually includes module performance data for a Nominal Operating Cell Temperature of 49°C, ambient temperature of 20°C, irradiance of 800 W/m², and average windspeed of 1 m/s. These are sometimes called Standard Operating Conditions. The higher operating cell temperature has a considerably degrading effect on module performance.

REFERENCES

[1]   D.M. Chapin, C.S. Fuller, and G.L. Pearson, "A new silicon p-n junction photocell for converting solar radiation into electrical power," *J. Appl. Phys.*, vol. 25, pp. 676-677, 1954.

[2]   *Electric Power Monthly*, Table 3, p. 9, Energy Information Administration, U.S. Department of Energy, May 1995.

[3]   Photovoltaic News, ed. by P.D. Maycock, vol. 15, p. 4, Feb. 1996.

[4]   M.A. Green and M.J. Keevers, "Optical Properties of Intrinsic Silicon at 300 K," *Progress in Photovoltaics*, vol. 3, pp. 189-192, 1995.

[5]   *Standard Tables for Terrestrial Direct Normal Solar Spectral Irradiance for Air Mass 1.5*, Spec. E891-87, American Society for Testing and Materials, Philadelphia, PA, Dec. 1987.

[6]   *Terrestrial Photovoltaic Measurement Procedures*, NASA Report TM 73702, June 1977.

2

SOLAR CELLS AS SEMICONDUCTOR DIODES

2.1 CRYSTALLINE MATERIALS AND SEMICONDUCTORS

Most pure solids exist in the form of a crystalline structure, or *lattice*, i.e., a regularly ordered array of atoms. If a pure material is allowed to freeze slowly from a melt, a regular atomic array is often the configuration of least free energy and is thermodynamically preferred. Thermal gradients and unavoidable impurities during the freezing process produce mechanical stress and dislocations in the crystal. These limit the long-range periodicity of the crystal.

A. Solids Exist in Crystalline Forms

Solids existing essentially as one crystal are referred to as *single-crystal*. They display periodicity throughout their bulk. A crystal may be many centimeters across. Frequently, they have a low density of imperfections. The crystal, of course, abruptly ends at the surface, but otherwise, it has few geometrical irregularities. Solids consisting of multiple crystals, but with each subcrystal being at least a tenth micron, or so, in size, are referred to as *polycrystalline*. When the individual subcrystals are very large (~ 0.1 cm across or larger), the polycrystalline material is sometimes called *semicrystalline* or *multicrystalline*. Collectively, single-crystal and polycrystalline materials are referred to as *crystalline materials*. Those solids with short range periodicity (perhaps only several nm) are referred to as *amorphous*. Oxides and glasses are examples of amorphous materials. These terms are frequently seen in solar cell product literature. Solar cells have been made from single-crystal, polycrystalline, and amorphous semiconductors.

The fundamental geometrical arrangement or unit of a crystal is the crystal's *unit cell*. The defining property of the unit cell is that it can be "translated", i.e., moved in multiples of exact x-, y-, and z-increments, to generate the entire lattice. For silicon and gallium arsenide, the increments x, y, and z are equal, so that the unit cells have the shape of a cube. The edge length of the cube is called the *lattice constant, a.*

A common unit cell is the "face-centered cubic" or *fcc*. This arrangement is a cube with one atom at each corner of the cube, and one atom at the center of each of the six faces of the cube (fig. 2.1-1). In pure single-crystal silicon, each atom is at the apex of a tetrahedron, equidistant from the four other atoms of the tetrahedron. This geometry is equivalent to that produced by two fcc unit cells that interpenetrate each other along their cubic diagonals by a distance equal to one-fourth of the diagonal length. It is the so-called "diamond" structure, and represents the *silicon unit cell*. Each unit cell has eight atoms with the distance between nearest neighbors being approximately 0.433a.

By comparison, gallium arsenide also displays the diamond structure, except that one of the interpenetrating fcc sublattices is composed of gallium atoms, and the other is composed of arsenic atoms. Each gallium atom has four nearest neighbors that are arsenic, and vice versa. The resulting unit cell is referred to as the *zincblende* structure. Thus, single crystals of silicon and gallium arsenide have the same lattice structure, except that in gallium arsenide, the individual sublattices are composed of gallium atoms and arsenic atoms, instead of just one species as in silicon (fig. 2.1-1).

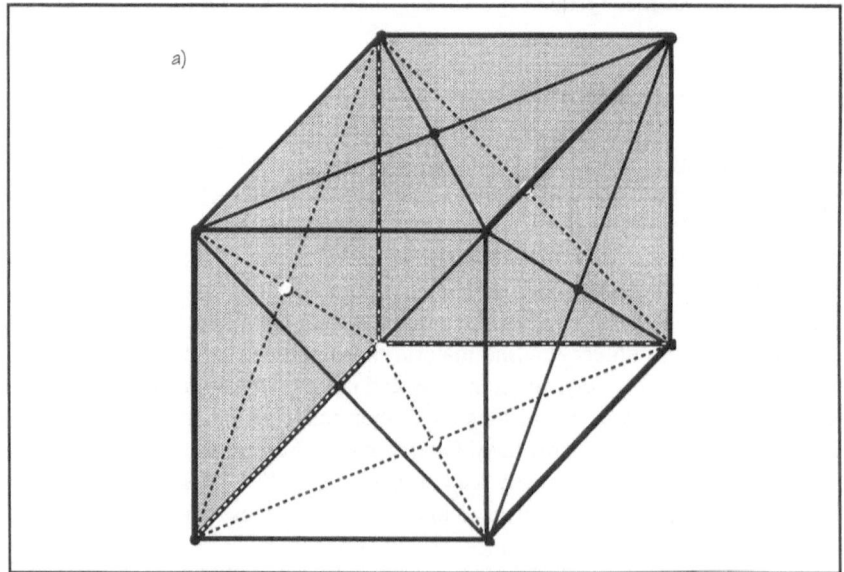

Fig. 2.1-1 (a)   Some common crystalline unit cells: (a) face-centered cubic, *above*; (b) diamond, and (c) zincblend lattice structures, *facing page*.

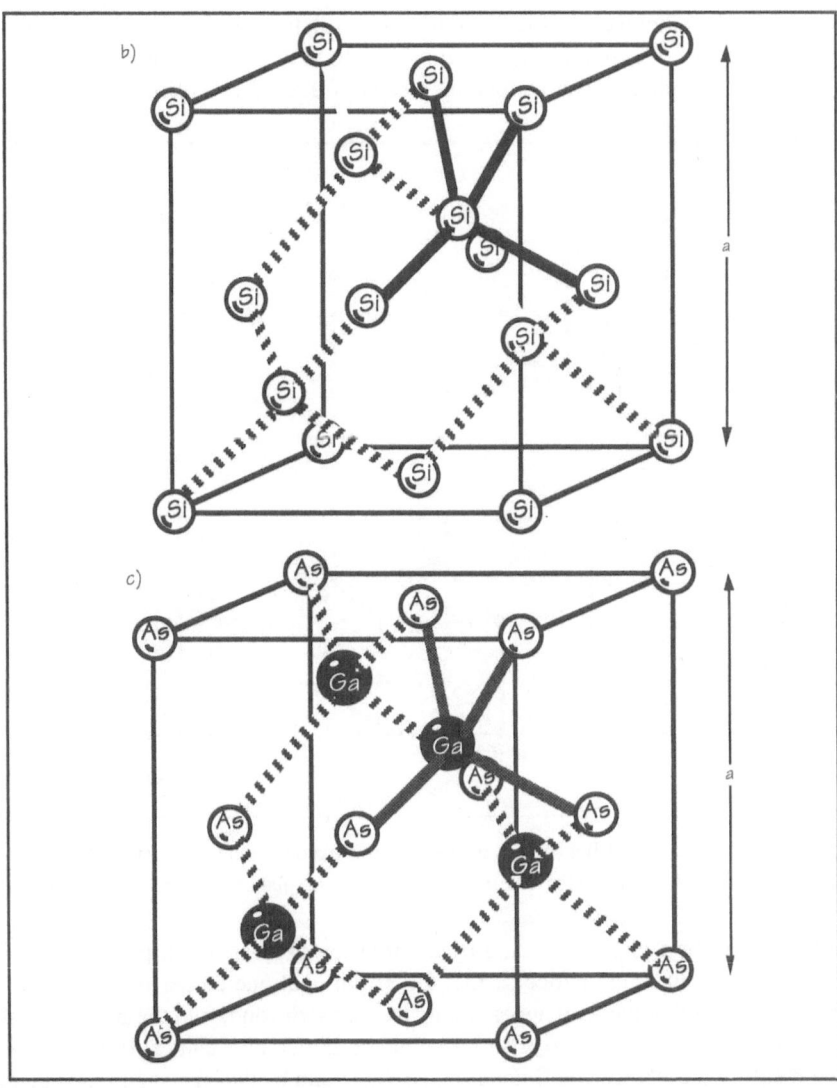

Diamond and zincblend structures do not have the same number of atoms per unit area in all planes. This is significant because planar density affects the electronic properties of the soild. Drift of electrons moving under the influence of an electric field varies with the planar atomic density. Also, the density of charge states that accumulate at the interface of the silicon crystal with a common insulator, silicon dioxide, is dependent on planar density. Consequently, it is important to have some method for distinguishing the different planes. The is done by means of the *Miller indices* (fig. 2.1-2). The

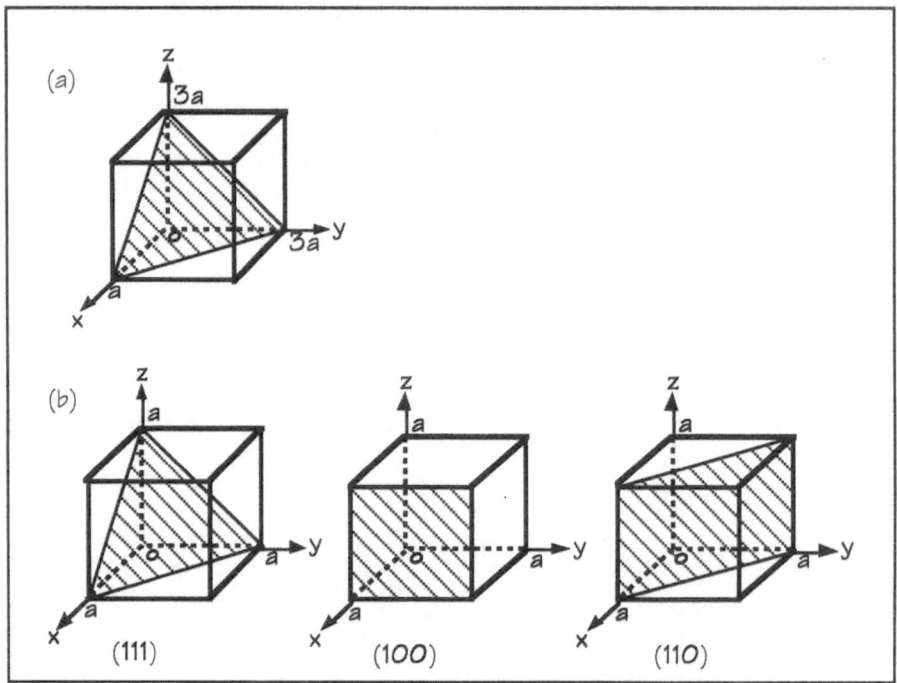

Fig. 2.1-2   The Miller indices distinguish the various planes in a crystal lattice. (a) Example of a (311) plane. (b) The three most important planes in the silicon lattice.

Miller indices of a plane are found by a three-step process. First, place an x-, y-, z-coordinate system in the lattice so that the origin is at one corner of a unit cell, and other corners of that unit cell appear along the axes. Second, find the intercepts of the given plane with the three axes in terms of the lattice constant, a. Finally, take the reciprocals of these three coefficients, and divide them by the largest common denominator. For example, if a plane, intersects the x-, y-, and z-axes at (4a,0,0), (0,4a,0), and (0,0,4a), respectively, then the intersection coefficients are 4, 4, and 4. The reciprocals are 1/4, 1/4, and 1/4. The largest common denominator is 1/4. Thus, this plane has Miller indices of 1, 1, and 1. This is written as (111). As another example, a plane parallel to the yz-plane and intersecting the x-axis at x=a would be labelled (100). For the diamond and zincblend structures, the (100), (010), and (001) planes are all equivalent. That is, they display equivalent symmetry. The set of planes equivalent to the (100) plane is referred to as {100}. The direction of a plane is the orthogonal to that plane. For example, the x-axis is the direction of the (100) plane. It is written as [100]. A set of equivalent directions to this direction is written as <100>. For the diamond and zincblende structures, {111} planes have a higher density of atoms per unit area than {100} planes. This difference is critical when orientation-dependent silicon etchants are used to texture the top surface of a solar cell to enhance light trapping. The surface is anisotropically textured with four-sided pyramids that reflect light into the surface rather than back into space.

The concept of a crystal completely free of defects is an abstraction. Every crystal will have some *point defects*. These include sites where a regular lattice atom is missing (vacancy), an atom between normal lattice sites (interstitial), two side-by-side vacancies (divacancy), a vacancy-interstitial pair (Frankel defect), and foreign atoms. A foreign atom can occur substitutionally on a regular lattice site or interstitially. For compound semiconductors, an atom of one species can occur on a site regularly occupied by the other species (antisite defect). Additionally, a crystal may have *dislocation* defects, in which groups of atoms are missing or misoriented. An example is an edge dislocations where a plane of atoms in the lattice suddenly terminates. Other defects include stacking faults where succeeding layers in the crystal are dissimilar, and twinning where regions of dissimilar crystallographic orientation share a common plane. Twins appear as straight lines on the surface of a wafer. Frequently, regions of dissimilar orientation will not share a common plane and the interface (*grain boundary*) appears as a jagged line on the wafer surface. Defect occurrance is strongly dependent on processing conditions. Thermal cycles with high temperature are associated with high defect densities. However, while some carefully grown crystals and subsequent solar cells are essentially free of dislocations, twins, and stacking faults, all crystals have an expected concentration of thermodynamically stable vacancies equal to $\exp(-E_a/kT)$, where E_a is the activation energy required to create a vacancy and k is Boltzmann's constant[1].

B. Electrons in a Crystal Reside in Energy Bands

From the quantum theory of solids, it is known that electrons in an isolated atom are restricted to certain discrete energy levels. When N atoms are brought into proximity of each other, as in a crystal, each energy level divides into N discrete but closely spaced levels to constitute a continuum or band of energies. Adjacent bands are separated by a range of energies which are forbidden to the electrons in a pure crystal (fig. 2.1-3).

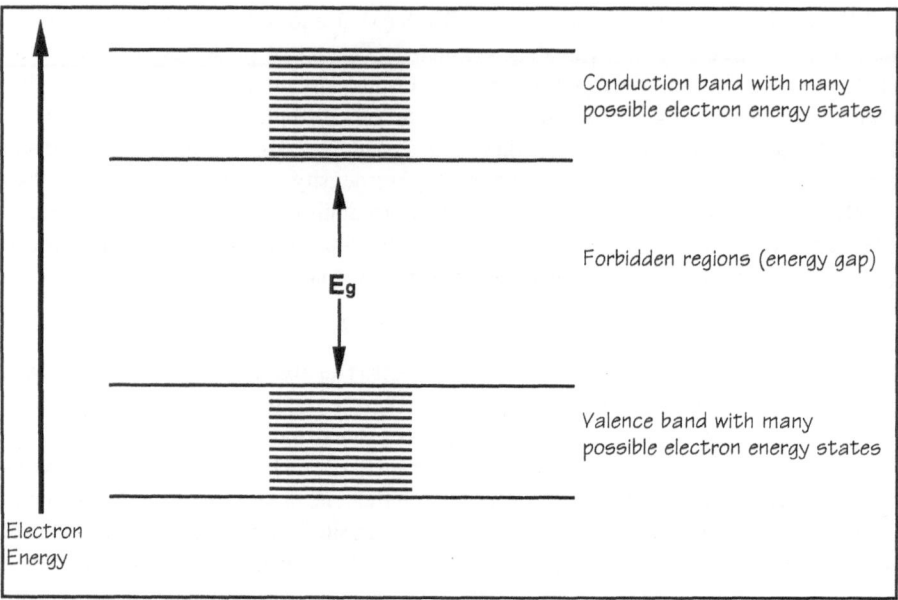

Fig. 2.1-3   Discrete energy bands in a crystalline solid.

Temperature is a measure of the median kinetic energy of an ensemble of particles in equilibrium with each other. Consequently, the absolute zero of temperature (zero Kelvin) represents the unachievable condition in which all kinetic energy in the crystal disappears. If the temperature of a crystal could be lowered to absolute zero, all of the energy would be potential energy. For this condition, the highest energy band with any electrons in it would be completely filled. This is the *valence band*. It is associated with chemical bonding. Valence electrons are tightly held by the periodic potential of the positive nuclei in the lattice. For temperatures above absolute zero, some of the electrons in the valence band increase their *potential energy* by jumping into energy states in the next highest energy band – the *conduction band*. Since an electron at the bottom of the conduction band is at rest, the movement of an electron from the top of the valence band to the bottom of the conduction band represents an increase in potential energy only. Such an electron has not changed in position; it has merely changed potential energy. At any given temperature, electrons will move back and forth between the valence and conduction bands in a random manner. For the crystal constrained to a fixed temperature, this is a statistical process with equal rates in both directions. The process maintains a certain time-average concentration of electrons in the conduction band which varies only with temperature. As the temperature increases, more and more electrons can make the transition, and the time-average number of electrons in the conduction band increases.

Electrons in the conduction band can gain kinetic energy. This can occur by the application of an electric field which imparts an acceleration to the electrons, or by other processes. When an electron gains kinetic energy, it moves above the bottom of the band. Thus, the incremental energy between the electron's level in the conduction band and the bottom of the band represents the electron's kinetic energy. Electrons tend to reside in energy states near the bottom of the band. If kinetic energy is imparted to an electron, it will quickly tend to lose that energy through lattice collisions so that it once again moves down toward the bottom of the band. The kinetic energy lost by the electron is converted to quantized lattice vibrations, i.e., *phonons*. This process whereby an energetic electron high up in the conduction band loses kinetic energy and moves toward the bottom of the band is called *thermalization*. It proves to be very important in determining the realizable efficiency of a solar cell. When an electron moves into the conduction band, the less energetic electrons in the lower energy bands shield it from the periodic attractive force of the positively charged atomic nuclei. This allows conduction band electrons to behave as essentially free, negatively-charged particles. They are free to *drift* under the influence of an externally applied electric field and form a current. The energy difference between the top of the valence band and the bottom of the conduction band is called the *energy gap*, or *bandgap* of the material, E_g. It is the single most important parameter of a crystalline (or amorphous) material. For a *pure* crystalline material, the energy gap constitutes a "forbidden" range of energies, where no electrons are allowed.

Some irregularities in the crystal periodicity constitute defect sites and introduce energy states in the bandgap which can be occupied by electrons. Irregularities that introduce bandgap states tend to lower solar cell performance. Other irregularities do not lead to states in the bandgap. Sometimes, these irregularities even have a net positive effect in that they neutralize bandgap states caused by other irregularities. An example is interstitial oxygen in certain polycrystalline ribbon solar cells where oxygen is suspected to passify a variety of defects[2]. Hydrogen in silicon tends to tie up bandgap states by bonding with uncoordinated (not fully bonded) silicon atoms, and is frequently used as a passivant. Another example is the preferential diffusion of phosphorus along grain boundaries in polycrystalline cells. In this case, phosphorus passifies the deleterious effects of non-fully bonded silicon atoms that appear along the grain boundary. Creation of defects can also beneficially affect the resistivity of a material. Many semiconductor devices require an ability to create high resistivity material, i.e., material with a low concentration of free charge carriers. In the preparation of certain crystals, e.g., gallium arsenide ingots, the unintentional incorporation of impurities (contaminants) can introduce an energy state near the conduction or valence band edge. This facilitates the occupancy of charge carriers in either the conduction band (electrons) or valence band (holes) and greatly decreases resistivity of the ingot. An example is carbon contamination in GaAs. Carbon creates an energy state at 0.019 eV above the top of the valence band. The state easily accepts electrons from the valence band, leaving holes (effective positive charges) in the valence band. This is compensated by the naturally occurring EL2 defect that is introduced by ingot growth in a pyrolytic boron nitride crucible. The EL2 defect is caused by the presence of an As atom on a Ga lattice site, perhaps also involving an adjacent vacancy. This defect introduces an energy state at 0.76 eV below the conduction band edge. Its deep position in the bandgap makes it

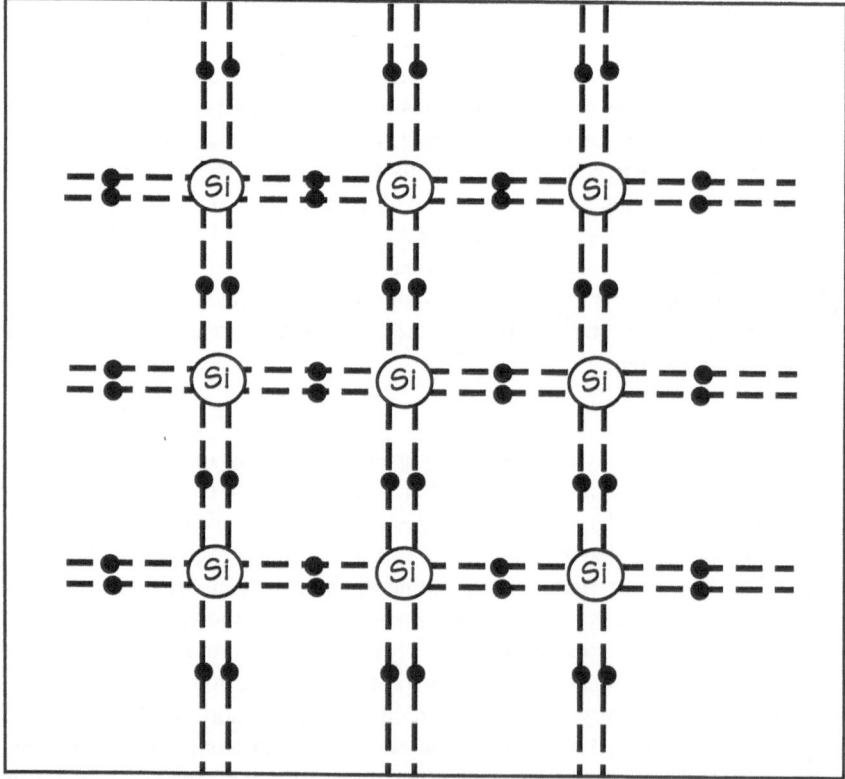

Fig. 2.1-4   Most semiconductors, including silicon and gallium arsenide, display covalent bonding. Each pair of atoms share two weakly-bonded electrons.

an efficient intermediate step for electrons that fall back into the valence band, thus neutralizing the effect of the carbon.

Like most semiconductors, silicon crystals display covalent bonding (fig. 2.1-4). In this scheme, a pair of atoms shares a pair of valence band electrons having equal energies but opposite quantum spins. The shared electrons are weakly bonded to each of the two atoms. In a pure defect-free silicon crystal near absolute zero, every atom has four valence electrons that it shares with its four nearest-neighbor silicon atoms. However, as the temperature increases, some of the bonds holding valence electrons are broken. The potential energy of the electron then exceeds the strength of the bond, and the electron moves from the valence band to the conduction band. Additionally, valence electrons that are not involved in bonding can be thermally excited so that they too jump into the conduction band. The thermal excitaton of an electron into the conduction band does not necessarily imply that a Si-Si bond has been broken. At room temperature, only a small fraction of valence electrons will have enough thermal energy to jump into the conduction band. For silicon at 300 K, only about 10^{10} electrons per cubic centimeter, or about 1 electron for every 10^{12} atoms, will have sufficient energy to do this. However, the very small addition of electrons to the conduction band (and corresponding losses in the valence band) has a drastic effect on resistivity. For this reason, thermally-induced electron energy transition is a critical mechanism in semiconductors, and leads to a special concept for describing empty states in the valence band.

C. Holes Represent Missing Electrons in the Valence Band

When an electron in the valence band jumps into the conduction band, it leaves behind an empty energy state, or *hole* (fig. 2.1-5). In the absence of impurities and defects,

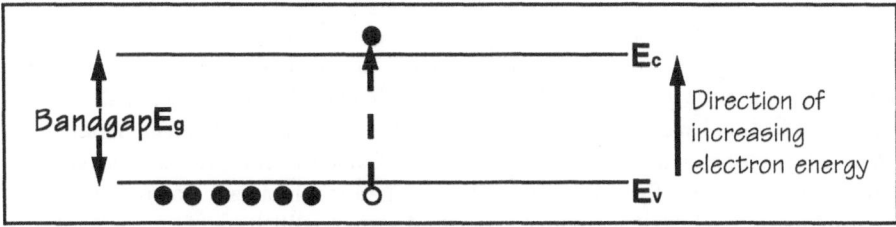

Fig. 2.1-5   When an electron in the valence band (not necessarily in a bonding pair) jumps into the conduction band, it leaves behind a hole, or effective positive charge, in the valence band.

conduction band electrons and valence band holes are created in *pairs* and their numbers are equal. The concept of a hole in the valence band can be thought of as a bookkeeping mechanism for keeping track of the empty energy states in the valence band. In this sense, the hole is an artifact created by an electron that has enough potential energy to jump out of the valence band. Whereas an electron is negatively charged, the hole behaves effectively as a *positively-charged particle*, with a charge magnitude equal to that of the electron. The validity of the concept of the hole as a positively-charged particle free to move about in the valence band is well established in semiconductor physics[3]. The farther down in the valence band that the electron starts from, the greater is the energy needed to elevate it to the conduction band. Thus, deep holes must represent more energy than shallow holes. For a hole, the direction of increasing energy is downward in the valence band. Analogous to an electron at the bottom of the conduction band, a hole at the top of the valence band has zero kinetic energy. If an external electric field is applied across the sample, the hole is accelerated and gains kinetic energy. It moves down in the valence band. Also analogous to energetic electrons in the conduction band, energetic holes tend to lose their kinetic energy (thermalization) and move toward the top of the valence band.

If there is any irregularity in the crystal, defect states are created in the forbidden zone. Defect states can serve as temporary energy states for electrons and affect the rate of transition between bands. In general, there are two mechanisms for electrons moving between bands. In a band-to-band transition, electron moves directly between bands. In a defect-state assisted transition, the electron makes an intermediate stop in the defect state. Intermediate states near the middle of the energy gap are the most efficient for aiding transitions. Defect states in the bandgap near the bottom of the conduction band easily capture or emit an electron with respect to the conduction band, but are much less likely to capture or emit an electron with respect to the valence band. An analogous statement is true for holes with respect to defect states near the top of the valence band. The most efficient state for assisting transitions is in between these two extremes, i.e., a state near the middle of the bandgap[4,5]. Regardless of the transition mechanism, the elevation of an electron from the valence band to the conduction band always results in the *generation* of an electron-hole pair. Conversely, when an electron in the conduction band falls back into a hole, either directly or through an intermediate state, a free electron and a hole simultaneously disappear. This process – *recombination* – is the opposite of pair generation. When a pair recombines, an amount of energy approximately equal to the bandgap is released, either in the form of a particle of light (photon) or a quantized vibration in the lattice (phonon). Emitted energy is expected to be just slightly larger than the bandgap due to the energetically favored location of holes and electrons near the respective band edges. Free electrons are most likely to reside near the bottom of the conduction band, while holes are most likely to reside near the top of the valence band.

Energy band structure determines which mode of recombination is dominant. In silicon and germanium, recombination is dominated by the existance of defect states in the bandgap. This is expected from the type of bandgap seen in these materials. Silicon and germanium are examples of *indirect bandgap* materials. For such crystals, a plot of electron energy verses electron crystal momentum shows the minimum of the conduction band and the maximum of the valence band at different crystal momentum values. This observation underscores the difference between completely free electrons and conduction band electrons. For a completely free electron, zero kinetic energy implies zero particle momentum. However, the periodic potential of the atomic nuclei causes conduction band electrons to behave as if they have an effective mass different from the mass of the free electron. A consequence from quantum mechanics is that the conduction band electron can have non-zero momentum even when the kinetic energy is zero. For any transition between bands, both energy and momentum must be conserved. This requirement makes band-to-band transitions in indirect bandgap materials very unlikely. To conserve both energy and crystal momentum, electronic transitions through defect states in the bandgap are much more favored than direct transitions between bands. These transitions involve a quantized lattice vaibration. On the other hand, *direct* bandgap materials have their maximum and minimum energies at the same value of crystal momentum (fig. 2.1-6). For these materials, both energy and momentum can be conserved by the electrons moving directly between bands. Transitions can also occur through defect states, but are not as energetically favored as in indirect bandgap materials. Accordingly, indirect bandgap materials tend to be more sensitive to crystal preparation and processing conditions, the prime determinators of defect concentration, than do direct bandgap materials. The differentiation between direct and indirect bandgap explains many of the photonic qualities of semiconductors. Such qualities as travelling charge domains in gallium arsenide diodes used for microwave circuits (Gunn diodes) and the unsuitability of silicon for light-emitting diodes are immediate consequences of the bandgap structure.

From the standpoint of solar cells, the main significance of bandgap type is that it determines the rate of absorption of sunlight and the dominant mechanism by which electrons recombine with holes. Indirect bandgap materials have much greater absorption lengths than direct bandgap materials because the probability (per unit length) of absorption is greatly decreased by the infrequency of band-to-band transitions. As an example, the absorption length at 0.80 µm (near infrared) in silicon is 11.8 µm, but only 0.6 µm in GaAs. Gallium arsenide cells can be fabricated from thin films because

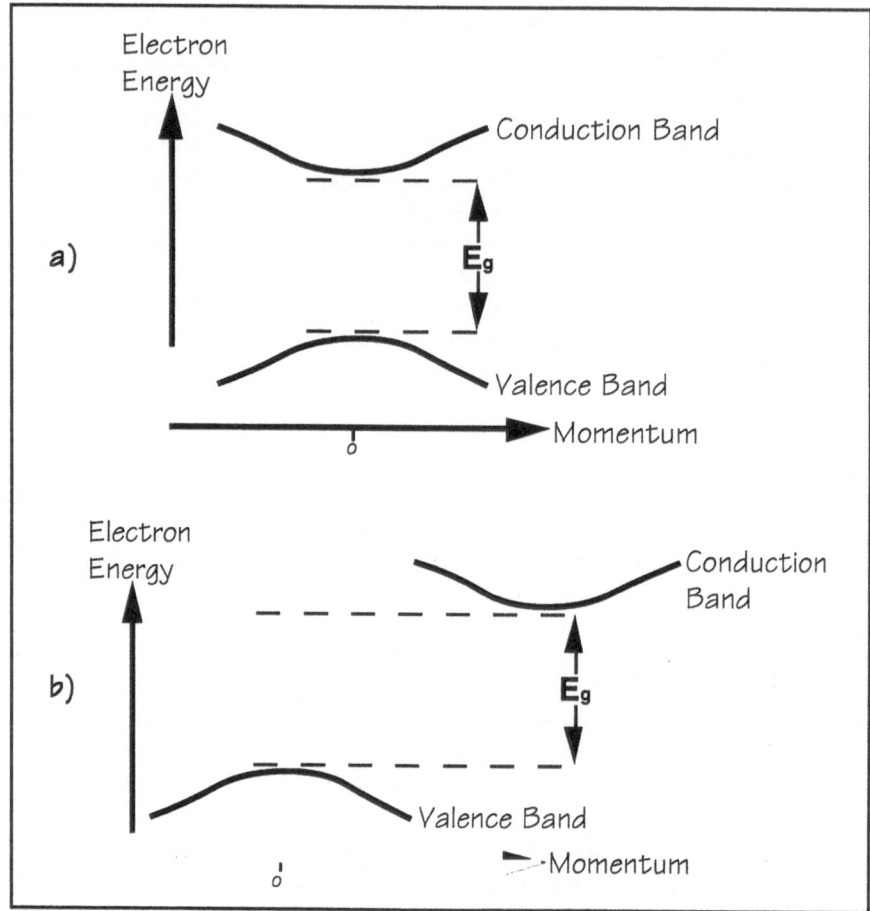

Fig. 2.1-6   Two types of bandgap: (a) direct, and (b) indirect, are distinguishable
by the conduction band minimum energy vs. crystal momentum.

of the high absorption coefficient. Silicon requires thick material for absorption (or
equivalently, multiple passes through the material.) This has implications for price and
specific power (watts/kg), particularly if a thin film can be deposited on a light, relatively inexpensive, substrate. A good example is GaAs on Ge.

Impurity atoms sometimes disrupt the regularity of the lattice and introduce defect
states in the bandgap. An impurity atom has an ionization energy associated with its

own valence electrons. Ionization refers to the gain or lose of a valence electron. If an impurity atom contributes an electron to the conduction band, it becomes positively ionized. If it takes an electron from the valence band, it becomes negatively ionized. In either case, the energy for the ionization can be compared to the energies of the conduction and valence band edges. For example, if it requires 0.1 eV to contribute an electron to the conduction band of the lattice, the positive ion and any defect can be thought of as residing 0.1 eV below the bottom of the conduction band. Likewise, if a lattice valence electron must gain 0.1 eV to become a valence electron of the impurity, the resulting negative impurity ion and any defect associated with it are thought of as lying 0.1 eV above the top of the lattice valence band. Because the recombination process in indirect bandgap materials like silicon is highly dependent on intermediate states in the bandgap, the position of an impurity ion and accompanying defect is very significant. Impurities with large ionization energies are not easily ionized. Instead, they introduce energy states near the middle of the bandgap and greatly enhance the rate of recombination (generation) of excess carriers in silicon. Impurities with small ionization energies have little effect on the rate of recombination, but greatly affect the concentrations of carriers. Thus gold in silicon, with an ionization energy 0.54 eV below the bottom of the conduction band, drastically increases the recombination rate of excess carriers[6]. On the other hand, arsenic, with an ionization energy 0.054 eV below the bottom of the conduction band, serves as an electron donor and does not participate in recombination.

Pure crystalline materials are classified by the size of their energy gaps. At room temperature (27° C or 300 K), if the energy gap is more than about 3 eV, very few electrons in the valence band have enough thermal energy to jump into the conduction band. There are few charge carriers to drift through the material when an electric field is applied. The material does not support much of an electric current. Such a material is an *insulator*. Silicon dioxide, SiO_2, ($E_g \sim 9$ eV) and silicon nitride Si_3N_4, ($E_g = 5$ eV), both widely used as surface defect passivants and anti-reflection coatings are good examples of insulators. For energy gaps between about 0.3 eV and 2.0 eV, many more electrons are able to make the jump into the conduction band, and the material is a *semiconductor*. Silicon ($E_g = 1.12$ eV), germanium ($E_g = 0.66$ eV), and gallium arsenide ($E_g = 1.42$ eV), are examples. If the energy gap is very small, or if the valence band overlaps the conduction band, the material has a huge concentration of electrons in the conduction band and supports a large current when a field is applied. This is a *conductor*. Pure metals are good examples of conductors. At a high enough temperature, any crystalline solid has some of the attributes of a metal because of the large concentration of electrons that have jumped up into the conduction band. Some insulators become conductive at high temperatures due to ionic conduction. Bandgap is the most important physical parameter of a *crystalline solid* because it determines whether the crystal is an insulator, semiconductor, or conductor. However, the distinction between insulators and semiconductors is more complicated than the simple picture provided by these definitions. Some materials with large bandgaps, e.g., ZnO, conduct well when doped with suitable impurities that add electrons to the conduction band. Other insulators, e.g., Si-suboxides, remain non-conducting even with impurity doping.

Semiconductors tend to differ from insulators and conductors in several characteristics aside from bandgap. The most important of these is the strong directional covalent bonding seen between nearest neighbor atoms in many semiconductors. Covalent bonding causes semiconductors to be hard and brittle, and to have negative coefficients of thermal expansion. They contract upon melting. Silicon and germanium are Group IV elements and have four electrons available for covalent bonding. Compound semiconductors include combinations of Group III and V elements, e.g., gallium arsenide and indium phosphide, and Group II and VI elements, e.g., cadmium telluride. The II-VI semiconductors have a strong ionic as well as covalent character. Additionally, there are several important tertiary semiconductor compounds, e.g., $Al_xGa_{1-x}As$, $In_xGa_{1-x}As$, and $Ga_xIn_{1-x}P_2$ in which the "x" indicates mole fraction. For tertiary semiconductors, mole fraction determines the nature of the bandgap and subsequent optical properties of the material.

In a pure covalent crystal, the concentration of electrons (holes) in the conduction (valence) band is called the *intrinsic concentration, n_i*. For any given semiconductor, the value of n_i is a function of temperature only:

$$n_i = \sqrt{N_C N_V} \; e^{-E_g/2kT} \qquad (2.1\text{-}1)$$

where k = 1.38 x 10^{-23} J/K is Boltzmann's constant, and N_C and N_V are the effective densities of states in the conduction and valence bands, respectively. Thermal generation of electron-hole pairs increases drastically with temperature, and decreases drastically with the magnitude of the bandgap. As an example, at 300 K, n_i = 1.5 x 10^{10}/cm^3 in silicon and 1.8 x 10^6/cm^3 in gallium arsenide. Solar cell conversion efficiency is dependent on intrinsic concentration because n_i is a determinator of the rate of recombination of excess electrons and holes during illumination – a parasitic mechanism in any solar cell.

Because electrons (in the conduction band) and holes are essentially free charged particles, they can move under the influence of an electric field. In general, in any physical environment, an electric field is created whenever there is a non-uniform distribution of charge, i.e., *separation of charge*. Electric field lines extend from the more positive region to the less positive region. Electric fields can exist in a silicon crystal either by an externally applied mechanism, e.g., a battery applied across the crystal (fig. 2.1-7),

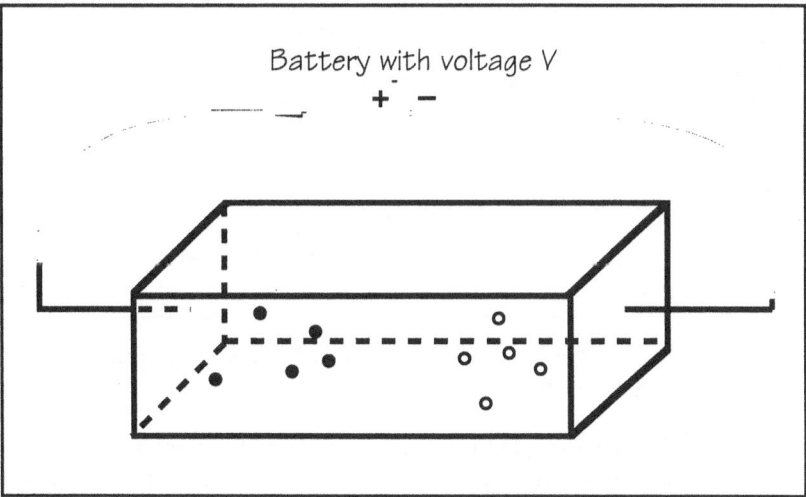

Fig. 2.1-7   The effect of a battery applied across a piece of semiconductor.
Electrons move toward the positive terminal; holes move toward the negative terminal.

or because of built-in non-uniformities in the distribution of the electrons and holes in the crystal. Electrons move against the field with an acceleration proportional to the field strength (Coulomb's Force Law). Collisions with the lattice and impurity ions cause the average velocity of the electrons to reach a uniform value. The speed per unit electric field is the *mobility* μ, and the electron is said to *drift* in the field. The magnitude of the drift velocity is much less than the individual electron speeds since the electrons experience large accelerations during their random walk through the lattice. Thus, the mobility characterizes the net drift of charge carriers that form a current. Mobility is dependent on the material, impurity concentration, and the temperature. A similar drift phenomenon occurs for holes in the valence band. However, because the polarity of the hole charge is positive, they will drift with the field. *Holes and electrons drift in opposite directions* (fig. 2.1-8). The effective masses of the holes and electrons are different because of different magnitudes in the periodic nuclear attraction they experience as they drift through the lattice.

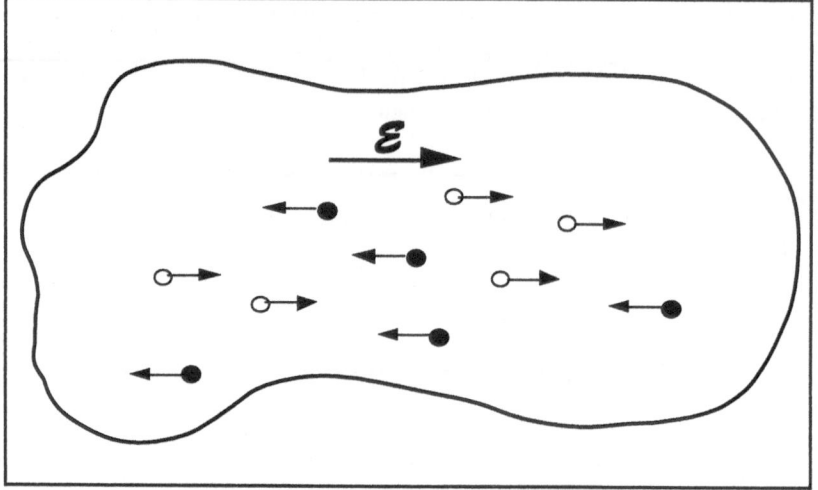

Fig. 2.1-8   Electrons and holes drifting in a semiconductor under the influence of an electric field.

Since the hole is a bookkeeping mechanism of sorts, it is important to have a good physical picture of how the hole moves in the valence band of a crystal under the influence of an electric field. When an external electric field exists in a crystal, the kinetic energy imparted to a valence electron by the field enables that electron to move into a nearby hole. This action results in a hole being created at that energy state (and physical position) where the migrating electron used to reside. By the same mechanism, a second electron then fills the new hole; again, only to leave behind another hole where the second electron used to reside. Electrons will drift against the field; the holes they leave behind will propagate in the opposite direction. This is consistent with the fact that positively charged particles drift with a field, and not against it. The process continues indefinitely. Holes drift in an electric field by this hopping action of valence electrons within the valence band (fig. 2.1-9). This helps to clarify why electrons (in the conduction band) and holes have different effective masses and mobilities. The effective mass of a hole is really the effective mass of a valence band electron. Kinetic energy that a valence electron gains when it is accelerated by an electric field is usually insufficient to elevate it to the conduction band. Thus, the application of a field across the sample does not produce holes. The concept of a hole is a convenient way to keep track of the behavior of electrons in the valence band as opposed to electrons in the conduction band. From now on, references to electrons will mean electrons in the conduction band unless otherwise noted.

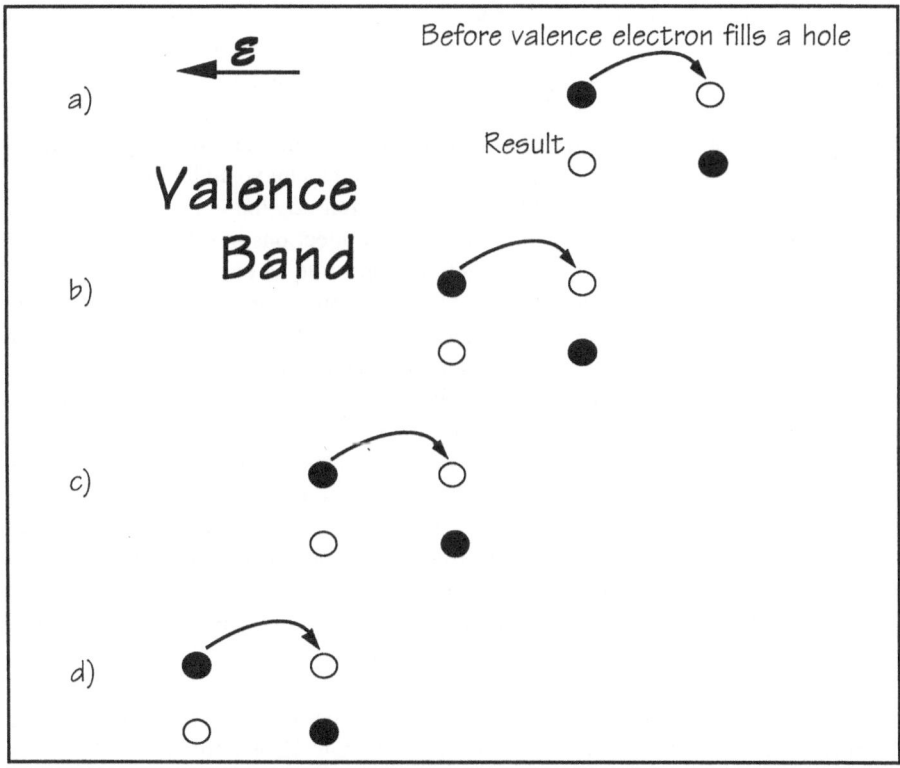

Fig. 2.1-9   A hole drifts through the valence band, under the influence of an electric field, by electrons hopping from energy state to energy state within the valence band.

D. Conductivity Depends on Band Occupation

For a homogeneous isotropic crystal, conductivity σ is a scalar that relates the current density vector to the electric field strength vector: $\mathbf{J} = \sigma \mathcal{E}$. Materials in general can be classified as insulators or conductors based on the magnitude of σ. However, crystalline materials are more precisely classified as an insulator, semiconductor, or conductor by the magnitude of the material's bandgap. This classification is useful when there are impurities in the material. It relates conductivity to the mobility and charge concentration produced by the presence of the impurity: $\sigma = q(\mu_n n + \mu_p p)$, where μ_n and μ_p are the electron and hole mobilities, respectively, and n and p are the electron and hole concentrations, respectively. Because different planes in a crystal have different atomic densities, the mobility, and subsequently the conductivity, are expected to be directionally dependent. This does not present a problem in most devices, including solar cells, because the current can often be approximated as one-dimensional.

For a crystalline material, if the valence band is almost full, there will be few holes in the valence band, and thus, few positive charges that can move under the influence of an applied electric field. There will be little hole current. By the same reasoning, if the conduction band is almost empty, there will be few electrons in the conduction band that can drift under an applied electric field to constitute an electron current. Such a material is effectively an insulator. As an example, a pure silicon crystal at room temperature behaves almost like an insulator, as it has only about 10^{10} electrons and holes per cubic centimeter – a very small concentration compared to the density of silicon atoms 5×10^{22} cm^{-3}. The valence band is almost full of electrons and the conduction band is almost empty. *On the other hand*, if the valence band can be partially emptied or the conduction band partially filled by some mechanism, e.g., by selectively introducing certain impurities into the lattice that contribute holes or electrons, one of the bands can take on a huge population. In this case, the valence band is selectively made partially empty or the conduction band partially full. The material will support either a large hole current or a large electron current, respectively, when an electric field is applied. Thus, by partially emptying the valence band (adding holes) or partially filling the conduction band, the crystal becomes a conductor.

This shows one of the most interesting and useful properties of semiconductors. Their bandgaps fall in a range of values that makes it easy to change the material from an insulator to a conductor by selectively introducing a small amount of impurity. (There are other mechanisms for making this change, for example, application of gate voltage in a MOS transistor.) Thus, a better terminology for semiconductors might be the term *"sometimes conductors"*. This is not the case with true insulators and conductors. An insulator, e.g., ceramic, never conducts much of a current. And a conductor, e.g., metal, always conducts a large current. The relative fullness of the valence and conduction bands in a semiconductor determines whether the crystal behaves like a conductor or an insulator, or something in between. *As long as at least one of the two bands has plenty of charges, the semiconductor is conductive.* If neither band has many charges, then the semiconductor is semi-insulating.

A mechanical analogy to this property can be made to a glass tube containing water and sealed at both ends. For such an object, two "species" can be identified: water and water-free pockets (or air bubbles). The water can be thought of as electrons, the air bubbles can be thought of as holes, and agitation of the tube can be thought of as an electric field. If the tube is completely filled with water, there is never any appreciable water current or air bubble current, no matter how much the tube is agitated. This corresponds to the case where the valence band is completely filled with electrons and the conduction band is void of electrons. But if the tube is only partly filled with water, then a little agitation will easily produce a current of water (electron current) and a current of air bubbles (hole current).

MATERIAL PROPERTIES: Si, GaAs		
	Si	GaAs
Atoms/cm^3	5.0×10^{22}	4.4×10^{22}
Density (g/cm^3)	2.33	5.32
Lattice Constant (nm)	0.5431	0.5653
n_i (cm^{-3})	1.5×10^{10}	1.8×10^6
Bandgap Type	Indirect	Direct
Bandgap (eV)	1.12	1.42
Electron Mobility (cm^2/V-s)	1450	8500
Hole Mobility (cm^2/V-s)	450	400
N_C (cm^{-3})	2.8×10^{19}	4.7×10^{17}
N_V (cm^{-3})	1.0×10^{19}	7.0×10^{18}

Table 1   Some important material properties of Si and GaAs at room temperature (300 K).

2.2 SOME SEMICONDUCTOR PROPERTIES

The introduction (or doping) of intentional impurities into a pure semiconductor crystal causes the sample to become either rich in electrons (negative or "*n-type*") or rich in holes (positive or "*p-type*"). As indicated in section 2.1D, the conductivity of a semiconductor crystal is greatly affected by changing the hole concentration in the valence band or the electron concentration in the conduction band. A remarkable property of semiconductors is that a small concentration of intentional impurity or *dopant* can have a drastic, though very predictable, effect on the conductivity of the crystal.

A. Conductivity Is Controlled by Introducing Impurities

It is often more convenient to refer to the reciprocal of conductivity (1 / conductivity) rather than the conductivity. This is *resistivity*, $\rho = 1/\sigma = 1/q(\mu_n n + \mu_p p)$, and has units of ohms-centimeters (Ω cm). It is a material parameter that reflects the concentration of the *charge carriers* (electrons or holes), and the scattering mechanisms that impede the drift of those charge carriers in an electric field. Resistivity does not depend on the geometry or shape of the sample under test. Metals and heavily-doped semiconductors have very low resistivities, typically 0.0001 to 0.01 Ω cm. Semiconductor material which is neither p-type nor n-type (*intrinsic* material) can have high resistivities, for example, 230,000 Ω cm for silicon and about 100,000,000 Ω cm for gallium arsenide, so that it is semi-insulating. Intrinsic germanium has a resistivity of only about 30-50 Ω cm. As an example of doping effect, if a sample of pure single-crystal silicon is doped with a small concentration of phosphorus, say one atom for every hundred thousand atoms of silicon, the resistivity of the crystal lattice decreases from 230,000 Ω cm to about 0.03 Ω cm. For such a sample, the corresponding current for a given applied voltage would increase by a factor of about 8,000,000.

Mobility μ varies with doping density and temperature and is dependent on mechanisms that scatter carriers[7]. These include scattering from the thermal vibrations of lattice atoms, scattering from the charge centers of impurity ions, and scattering by interactions between electrons and holes. Lattice and ionized impurity scattering are the most significant. For pure silicon, there is no impurity scattering and the mobility has a maximum. Mobility falls slowly as the impurity concentration is increased. Lattice scattering and ionized impurity scattering increase and decrease, respectively, with temperature. Thus, the mobility of silicon increases up to about 150 K, and then starts to decrease. Experimentally, for intrinsic silicon, $\mu \propto T^{-2.5}$ in the lattice scattering region[8]. However, the carrier concentrations increase quickly with temperature and this causes the resistivity to fall with temperature. This is one of the most striking properties of semiconductors, as opposed to metals: semiconductors have a *negative* coefficient of resistivity[9]. Eventually, at a high enough temperature, the creation of pairs dominates the majority carrier concentration, and the material becomes approximately intrinsic. In metals, on the other hand, the valence band overlaps the conduction band, all metal atoms are ionized, and the electron concentration equals the atomic concentration. Thus, with increasing temperature, there are no additional electrons in the conduction band to offset the degradation to μ from lattice scattering, and the resistivity in the metal increases.

A related parameter, *sheet resistance*, ρ_s, represents the resistance of a thin layer of

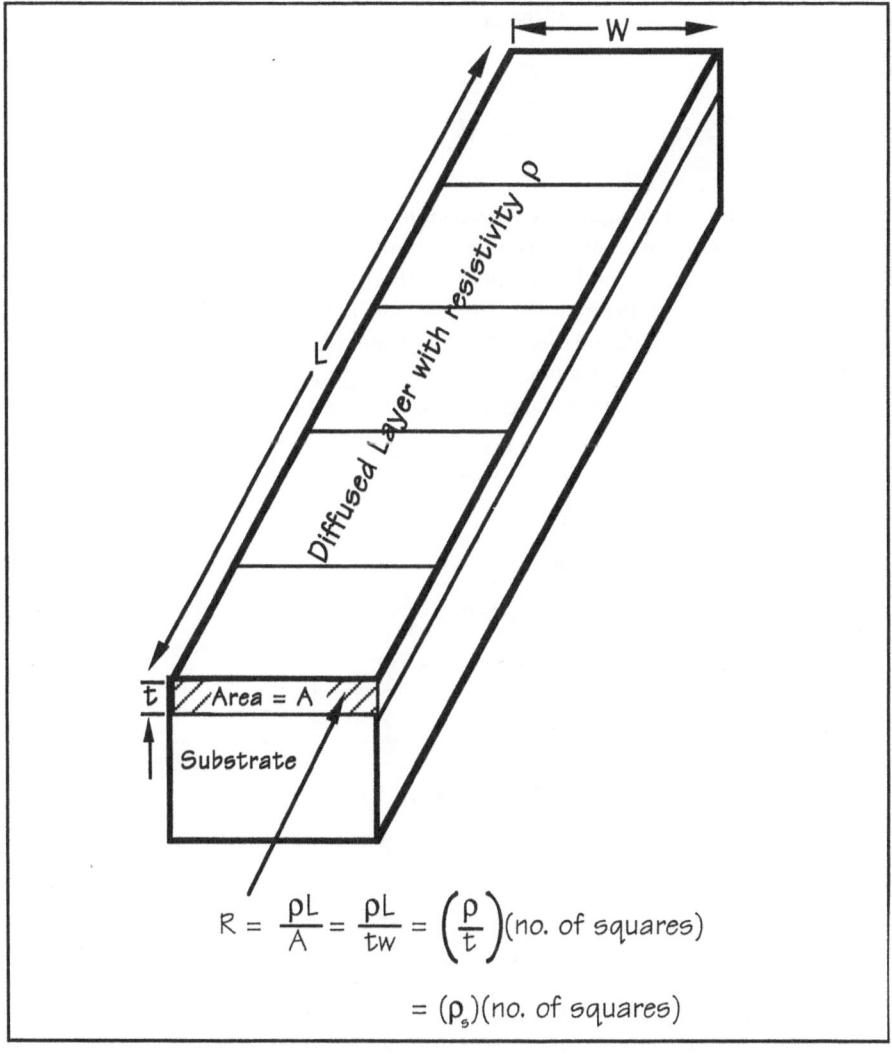

$$R = \frac{\rho L}{A} = \frac{\rho L}{tw} = \left(\frac{\rho}{t}\right)(\text{no. of squares})$$

$$= (\rho_s)(\text{no. of squares})$$

Fig. 2.2-1   Sheet resistance, ρ_s, is a measure of the resistance of a thin layer of material looking edge-on into the layer.

material (e.g., a layer of diffused impurity) looking edge-on into the layer. It is defined as $\rho_s = \rho/t$, where t is the thickness of the layer. From fig. 2.2-1, the resistance of the layer is $R = \rho L/A$, where L is the length and A is the cross-sectional area. Since

$A = WL$, where W is the width of the layer, the resistance may be written as

$$R = \frac{\rho}{t} \frac{L}{W} = \rho_s \,\square \qquad\qquad (2.2\text{-}1)$$

where $\square$ is the number of squares of material looking down on the layer. Sheet resistance has the units of ohms, which is often referred to as "ohms per square" for clarity. This is written $\Omega/\square$. If a conducting layer is 10 cm long and 2 cm wide, there are 5 squares looking down on the material. The resistance of the layer, looking into the 2 cm edge, is then $5\rho_s$. For a diffused n-type layer with an average phosphorus concentration of 10^{18} cm^{-3}, the average resistivity is 0.015 Ω cm. If the layer is 0.5-μm thick, the sheet rho is 300 $\Omega/\square$, and the total resistance looking into the 2-cm edge is 1500 Ω. Whereas

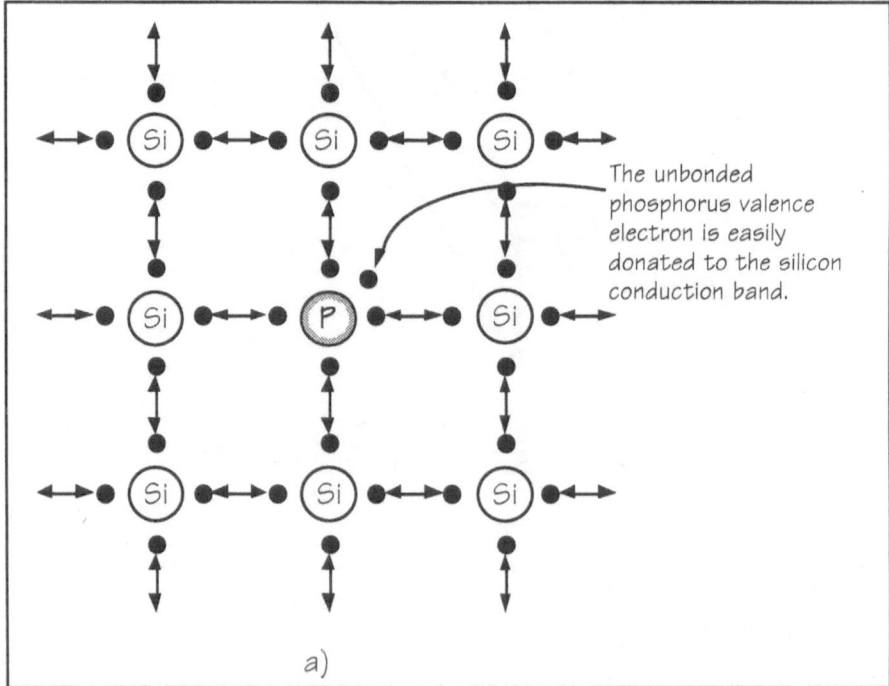

The unbonded phosphorus valence electron is easily donated to the silicon conduction band.

a)

Fig. 2.2-2   Various impurities can substitute for silicon atoms in the silicon lattice. (a) Phosphorus or arsenic increases the concentration of conduction band electrons in the

resistivity is strictly a material property, sheet resistance is a function of both material and geometry. Note that the layer does not have to be uniform (depth-wise) in its resistivity for sheet resistance to be defined. In almost all practical situations, resistivity of a diffused layer varies with depth.

Doping changes the resistivity of the semiconductor by adding holes or electrons to the semiconductor lattice as the dopant atoms are thermally ionized. A common example is phosphorus trichloride oxide (POCl$_3$) vapor and O$_2$ flowing through a quartz furnace tube containing silicon wafers. At 900°C, POCl$_3$ and O$_2$ readily form a phosphosilicate glass on the surface of the wafer. Immediately, phosphorus is reduced from the glass and diffuses into the silicon lattice. The phosphorus goes into the silicon lattice substitutionally. Each phosphorus atom forms four covalent bonds with its four nearest-neighbor silicon atoms, just as if it were another silicon atom (fig. 2.2-2a). However, phos-

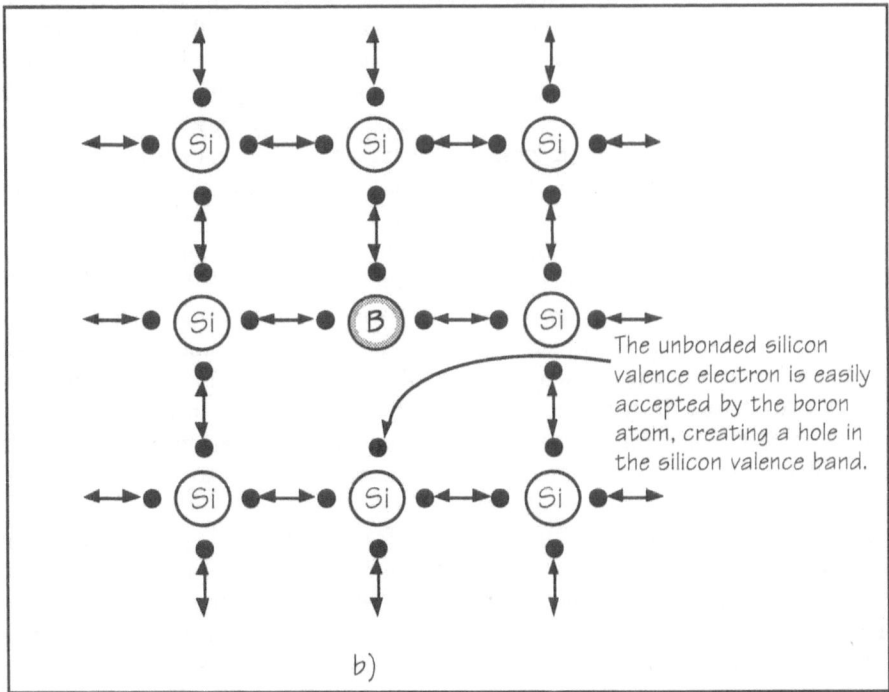

The unbonded silicon valence electron is easily accepted by the boron atom, creating a hole in the silicon valence band.

b)

silicon lattice. The phosphorus or arsenic atom becomes a positive ion. (b) Boron increases the concentration of holes in the silicon lattice. The boron atom becomes a negative ion.

phorus is in the fifth column of the periodic table, and has five valence-band electrons per atom that are available for forming covalent bonds. Silicon is in the fourth column of the periodic table, and has only four valence electrons per atom. The fifth valence electron in a phosphorus atom is left without a silicon atom with which to share a covalent bond. Because the ionization energy of the phosphorus atom (i.e., the energy necessary to free the electron from the phosphorus valence band) is about 0.045 eV below the bottom of the silicon conduction band, there is enough thermal energy available for the fifth phosphorus electron to jump into the conduction band of the silicon lattice. By this mechanism, an extra electron has been added to the silicon conduction band *without* creating a hole in the silicon valence band. The silicon has been made n-type. Because the phosphorus atom has donated an electron to the silicon lattice, phosphorus is referred to as a *donor* dopant. In n-type material, electrons are the *majority* carrier; holes are the *minority* carrier.

When the $POCl_3$ gas is turned off and the lattice is allowed to cool back to room temperature, the positively ionized phosphorus atoms are locked into the lattice. They are immobile below about 800°C. (It is assumed the sample is removed from the furnace and quickly cools down to room temperature so that the phosphorus does not have a chance to outgas.) High furnace temperature is not necessary for ionization. It serves only to create the phosphosilicate glass and enhance the subsequent rate of phosphorus diffusion into the silicon substrate. For temperatures above about 100 K (-173°C), most of the phosphorus atoms remain ionized, and the donated electrons are free to drift through the lattice in the conduction band. To maintain charge neutrality, it is necessary that $p + N_{DD} = n$, where n_0 is the initial concentration and where N_{DD} is the ionized impurity concentration. Total electron concentration is $n = n_0 + \Delta n$, and Δn is the excess electron concentration. For diffusions into p-type material, the concentration of conduction band electrons is approximately equal to the phosphorus concentration diffused into the lattice. Because the solid solubility of phosphorus[10] in silicon is 7×10^{20} cm^{-3} at 900°C, total electron concentration near the surface can be ten orders of magnitude greater than the intrinsic concentration when the sample cools down to room temperature.

By an analogous process, the diffusion of boron from a borosilicate glass into a silicon wafer forms a lattice rich in holes, i.e., *p-type silicon*. Boron has only three valence electrons per atom and forms only three covalent bonds with its four nearest-neighbor silicon atoms. Of the four nearest-neighbor silicon atoms, one of them is left with a valence electron that does not have the opportunity for coordination (fig. 2.2-2b). Because the ionization energy of boron (i.e., the energy necessary to capture a silicon valence electron) is about 0.045 eV above the top of the silicon valance band, there is enough thermal energy available for the extra silicon valence electron to become tightly held by the boron atom. The boron atom becomes negatively ionized. Absence of the valence band electron from the silicon lattice leaves the lattice with a *hole*. By this mechanism, an extra hole has been created in the valence band without adding an electron to the conduction band. Since boron has accepted a valence electron from the silicon lattice, it is an *acceptor* dopant. Phosphorus and boron are the standard n- and p-type dopants, respectively, because they have high solid solubilities and can render thin highly-doped diffused layers. Arsenic is also a donor in silicon. Though its solid solubility is larger than that of phosphorus, its use as a dopant is restricted to arsine-based ion implantation for integrated circuit fabrication.

There is a special relationship that exists between the hole and electron concentrations when the semiconductor is in *thermal equilibrium*. Thermal equilibrium means that both holes and electrons are in a condition of *detailed balance*, i.e., every hole current and every electron current is balanced by an equal and opposite hole or electron current, respectively. Carriers are constantly being generated and are constantly recombining; however, the net hole current and the net electron current are both zero in detailed balance. A silicon wafer held at a constant temperature in the absence of illumination and external electric fields is in a state of thermal equilibrium. Conversely, a solar cell exposed to sunlight is definitely not in thermal equilibrium. Thermal equilibrium is a special case of steady state. Steady state means that rates are constant, e.g., a constant current passing through a conductor. Thermal equilibrium is much more severe than steady state. It means that all net rates are zero. An analogy can be made to a river with a constant flow. While the number of water molecules passing per unit time per unit cross-sectional area is constant, it is not the case that all net rates are zero. Thus, the river is in steady-state, but not thermal equilibrium. To be in thermal equilibrium, the river would have to stop flowing, have no sources or sinks, and maintain a uniform temperature throughout its volume.

In thermal equilibrium, the product of the hole and electron concentrations is a constant dependent only on the temperature of the sample. This is the *Law of Mass Action* for semiconductors, and the constant is the square of the intrinsic concentration:

$$n \ p = n_i^2 \qquad (2.2\text{-}2)$$

where n_i is the intrinsic concentration, and n and p are the concentrations of electrons and holes, respectively.

As an example, consider a sample of silicon in thermal equilibrium and doped n-type with 10^{16} atoms/cm^3 phosphorus. Assume all the phosphorus is ionized. *Charge neu-*

trality requires, $n = N_{DD} + p$, where N_{DD} is the concentration of the positively ionized phosphorus donor. Since N_{DD} is so much greater than the intrinsic value of holes ($n_i = 1.5 \times 10^{10}/cm^3$ at 300 K), the electron concentration is approximately N_{DD} and $p = n_i^2/n = 2.3 \times 10^4/cm^3$. For a sample in thermal equilibrium, both the majority and minority carrier concentrations are fixed by the doping.

B. Energy State Occupation Probability

The introduction of easily ionized impurities into intrinsic semiconductor clearly affects the carrier concentrations. This is reflected in the probability of energy state occupation. For classical particles, e.g., gas molecules, the particles are identical and distinguishable, and the number of particles that can occupy an energy level is unrestricted. Such particles obey Maxwell-Boltzmann statistics. The probability that a classical particle has energy E at temperature T is $P(E) = C \exp[-(E-E_F)/kT]$, where C is a constant and E_F is the ground energy. For two energy states E_A and E_B, the ratio of the probabilities of occupation is $P(E_B)/P(E_A) = \exp[(E_A-E_B)/kT]$.

Electrons in a crystal, on the other hand, are identical and indistinguishable, and obey the Pauli exclusion principle that no two electrons can have the same set of quantum numbers, i.e., the same energy state. Thus, at most two electrons with opposite quantum spin can occupy the same energy level. If a level contains two electrons, it is said to be degenerate. The probability of a state with energy E being occupied by an electron is given by the Fermi-Dirac distribution

$$F(E) = \frac{1}{1 + e^{(E-E_F)/kT}} \qquad (2.2\text{-}3)$$

Likewise, the probability that a state (in the valence band) is occupied by a hole is $1 - F(E)$, and is given by

$$1 - F(E) = \frac{1}{1 + e^{(E_F-E)/kT}}$$

Energy E_F is called the Fermi energy and represents the combination of the electrostatic and chemical potentials for the electron, i.e., the electrochemical potential[11]. In thermal equilibrium, the probability that a state with this energy is occupied by an electron is

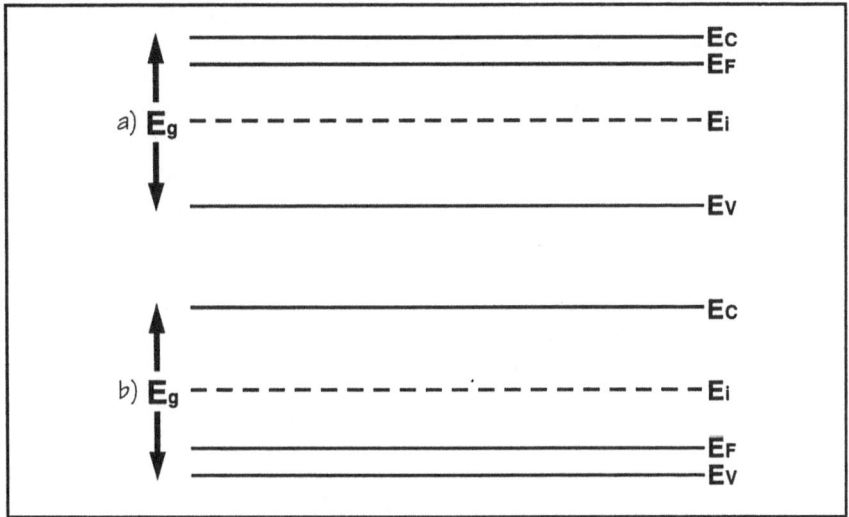

Fig. 2.2-3   Fermi level in (a) uniformly doped n-type material, and (b) uniformly doped p-type material. E_i is the Fermi level for intrinsic material.

equal to 1/2. The Fermi energy in intrinsic material is denoted as E_i, and is located very close to the middle of the bandgap. If intrinsic material is doped n-type, the Fermi energy is forced toward the conduction band. Likewise, if the material becomes rich in holes, the Fermi energy is forced toward the valence band (fig. 2.2-3, where energy is plotted as a function of position in uniform material). For a metal, E_F is above the bottom of the conduction band E_c. (This can be taken as the definition of a metal.) When an extrinsic, i.e., doped, semiconductor is heated, electron-hole pairs are created, and the material becomes more and more intrinsic as the number of pairs dominates the majority carrier concentration. For a high enough temperature, the Fermi energy in an extrinsic semiconductor will approach the middle of the bandgap, i.e., E_F approaches E_i at high temperature. Thus, the Fermi energy is a measure of the relative densities of electrons and holes. For n-type material, if E_c-E_F is more than 2kT (about 0.052 eV at T=300 K), then F(E) is approximately equal to $exp(E_F$-E)/kT, and the Fermi probability distribution apporaches the classical distribution. Otherwise, Fermi statistics must be used to determine distributions and the material is said to be *degenerate*. Physically, this means that there are so many electrons in the conduction band that some of the levels in the conduction band are taking on more than one electron and the exclusion principle becomes relevant. The levels can be occupied by at most two electrons. If there is only one electron per level in the conduction band, as in lightly n-doped material, then the issue of the exclusion principle does not arise and the distribution appears classical. Likewise, for p-type material, if E_F-E_V > 2kT, then the Fermi expression for the probability of hole occupation in the valence band is approximately equal to exp(E-E_F)/kT, and the distribution approaches the classical distribution. These approximations are highly useful because they simplify the expressions for electron and hole densities. Because the Fermi energy represents the electrochemical potential, a *fundamental conclusion* of quantum statistics is that the Fermi energy is constant throughout a device in thermal equilibrium.

Various expressions for the electron and hole densities in thermal equilibrium can be derived using the classical approximation for the Fermi distribution. These expressions assume the material is *non-degenerate*. For thermal equilibrium

$$n_O = N_C \, e^{(E_F - E_C)/kT}$$

$$(2.2\text{-}4)$$

$$p_O = N_V \, e^{(E_V - E_F)/kT}$$

where N_C and N_V are the densities of states in the conduction and valence bands, respectively, and E_C and E_V are the conduction and valence band edges, respectively. Here, the subscript "o" denotes thermal equilibrium concentrations. By noting that $E_g = E_C - E_V$, the product of equations in (2.2-4) leads to equation (2.1-2). In the case of intrinsic material, $n_o = p_o = n_i$, and $E_F = E_i$. Then

$$n_i = N_C \, e^{(E_i - E_C)/kT}$$

$$(2.2\text{-}5)$$

$$n_i = N_V \, e^{(E_V - E_i)/kT}$$

Solving equations (2.2-5) for N_C and N_V in terms of n_i and the exponentials, and substituting into equations (2.2-4) yields

$$n_O = n_i \, e^{(E_F - E_i)/kT}$$

$$(2.2\text{-}6)$$

$$p_O = n_i \, e^{(E_i - E_F)/kT}$$

This last pair of equations shows how the electron concentration increases quickly as the Fermi energy rises above the intrinsic level – which is approximately at the middle of the bandgap. Likewise, the hole concentration rises quickly as the Fermi energy falls below the intrinsic level. By noting that electron energy and potential are related by

$E = -q\phi$, the equilibrium hole and electron concentrations as a function of position can be written as

$$p_o(x) = n_i\, e^{-q\phi(x)/kT}$$

$$n_o(x) = N_V\, e^{q\phi(x)/kT}$$

(2.2-7)

where potential ϕ is taken as zero in intrinsic material in thermal equilibrium. These are the *Boltzmann* relations for holes and electrons. The minus sign in the equation for holes is consistent with the tendency of holes to move away from regions of high electric potential.

For non-thermal equilibrim conditions, the Fermi energy is replaced by *quasi Fermi* energies

$$n = n_i\, e^{(E_{FN}-E_i)/kT}$$

$$p = n_i\, e^{(E_i-E_{FP})/kT}$$

(2.2-8)

where E_{FN} and E_{FP} are the quasi Fermi energies for electrons and holes, respectively. For the non-thermal equilibrium case, $E_{FN} \neq E_{FP}$, and equations (2.2-8) can be written as

$$n = n_i\, e^{(qV_i-qV_N)/kT}$$

$$p = n_i\, e^{(qV_P-qV_i)/kT}$$

(2.2-9)

where V_N and V_P are the quasi Fermi potentials for electrons and holes, respectively, and $V_i = E_i/(-q)$ is the electrostatic potential for intrinsic material in thermal equilibrium and is taken as zero.

C. Hole and Electron Currents

In the bulk of a semiconductor, holes and electrons move by drift and diffusion. If the carriers are exposed to an electric field, for example, by placing the semiconductor in a circuit with a battery, the holes will drift with the field and the electrons against the field. This is a reflection of the Coulomb Force Law for a charged particle in an electric field, $\mathbf{F} = Q\mathcal{E}$, where Q is the algebraic value of the charge and $\mathcal{E}$ is the electric field strength vector. The hole and electron drift currrent densities are then

$$J_p = q\mu_p p \mathcal{E} \text{ and}$$

$$J_n = -q\mu_n n(-\mathcal{E}) = q\mu_n n \mathcal{E}$$

(2.2-10)

respectively, where the subscripts p and n refer to holes and electrons, q is the magnitude of the charge on the electron (1.602×10^{-19} Coulomb), and μ is the mobility. Current density J has units of amps/cm², electric field E has units of volts/cm, and mobility μ has units of cm²/volt-sec.

Holes and electrons also move in response to a concentration gradient, i.e., a spatial nonuniformity in concentration. In this case, the carriers will *diffuse* in the direction where the concentration is least. Carriers move away from regions of greater concentration to regions of lesser concentration. This phenomenon is similar to the diffusion of an odor from a bottle of solvent – as soon as the cap on the bottle is removed, the solvent molecules show a *net* movement in the direction of least concentration and spread throughout the room. The greater the nonuniformity of the hole or electron concentration, the greater will be the resulting hole or electron diffusion current. The hole and electron diffusion current densities are given by

$$J_p = q D_p(-\nabla p) = -q D_p \nabla p \text{ and}$$

$$J_n = -q D_n(-\nabla n) = q D_n \nabla n$$

(2.2-11)

where D is diffusivity and the minus sign in front of the gradient indicates carriers diffuse in the direction of least concentration.

From eqs. (2.2-10) and (2.2-11), the total hole and electron current densities are given by

$$J_p = q\mu_p p \mathcal{E} - q D_p \nabla p \text{ and}$$

$$J_n = q\mu_n n \mathcal{E} + q D_n \nabla n$$

(2.2-12)

respectively. In the one-dimensional case with carriers moving orthogonal to the solar cell surface, the gradient operator ∇ reduces to $(d/dx)\mathbf{a}_x$, where $\mathbf{a}_x$ is the unit vector. For the special case of zero current, the electric field can be expressed as

$$\mathcal{E} = D_p \nabla p / \mu_p p = -D_n \nabla n / \mu_n n \qquad (2.2-13)$$

It is interesting to note that diffusivity and mobility are related by a constant dependent only on temperature. This is seen by substituting the Boltzmann relation for holes and the magnetostatic expression for the electric field, $\mathcal{E} = -\nabla\phi$, into the equation for the total hole current at thermal equilibrium

$$J_p = 0 = q\mu_p p\mathcal{E} - qD_p\nabla p$$

$$0 = q\,\mu_p\,(n_i e^{-q\phi(x)/kT})\,(-\nabla\phi) - qD_p\nabla(n_i e^{-q\phi(x)/kT})$$

$$0 = q\,n_i e^{-q\phi(x)/k}\,[(\mu_p)\,(-\nabla\phi\,(x)) - (D_p)\,(-q\nabla\phi\,(x)/kT)]$$

The exponential prevents the factor $qn_i\exp[-q\phi/kT]$ from being zero. Thus,

$$0 = (\mu_p)(-\nabla\phi) - (D_p)(-q\nabla\phi/kT)$$

For the non-trivial case where $p(x)$ is not uniform, both sides can be divided by $\nabla\phi$, and μ_p and D_p are related by

$$D_p/\mu_p = kT/q \qquad\qquad (2.2\text{-}14)$$

known as the *Einstein relation*. There is a similar relationship for electrons.

For the non-equilibrium case, the total hole and electron currents can be expressed in terms of the quasi Fermi potentials. Here, it is assumed that holes and electrons move only by two mechanisms: drift and diffusion. Consider the holes. The velocity due to drift alone is

$$U_{drift} = \mu_p\mathcal{E}$$

$$= -\mu_p\nabla\phi$$

From statistical mechanics, the velocity due to diffusion alone is

$$U_{diff} = D_p\,\nabla[\ln(p/n_i)] \quad,$$

where ln is the natural logarithm. By using the Einstein relation

$$U_{diff} = D_p\,\nabla[\ln(p/n_i)]$$

$$= \mu_p\,(kT/q)\,\nabla[\ln(p/n_i)]$$

$$= \mu_p\,\nabla[(kT/q)\,\ln(p/n_i)]$$

Thus, ø is the potential associated with drift (i.e., electrostatic potential) and (kT/q)ln(p/n$_i$) is the potential associated with diffusion (i.e., chemical potential). The total velocity of holes is then

$$U_{total} = U_{drift} + U_{diff}$$
$$= \mu_p \, \nabla \, [\phi + (kT/q) \, \ln(p/n_i)]$$

This can be written as

$$U_{total} = \mu_p \, \nabla \, V_P$$

where $v_p = \phi + (kT/q) \ln(p/n_i)$ is the *electrochemical potential* for holes, which has already been identified as the quasi Fermi potential for holes. With the assignment of $\phi = 0$ in intrinsic material in thermal equilibrium, it is then straight forward to show that eq (2.2-12) for holes reduces to

$$J_p = -q\mu_p \, p \, \nabla \, V_P \qquad\qquad (2.2\text{-}15)$$

allowing the hole current to be expressed in terms of the quasi Fermi potential[12]. A similar expression relates the total electron current to the quasi Fermi potential for electrons. Note that for the equilibrium case, the quasi Fermi potentials reduce to the Fermi potential $E_F/(-q)$. Since the Fermi energy is flat in thermal equilibrium, the gradient is zero, and the total currents are then each zero, as expected.

2.3 PN-JUNCTIONS

When n-type semiconductor is brought into close contact with p-type semiconductor, a *pn-junction* or *pn-junction diode* is formed (fig. 2.3-1). This is the building block for bipolar transistors and almost all solar cells. A common procedure for creating this structure is the diffusion of a thin heavily-doped donor layer into the top side of a lightly-doped p-type substrate.

A. ENERGY BAND DIAGRAM AND BUILT-IN VOLTAGE

A useful tool for understanding the distribution and currents of charge carriers in semiconductor devices is the *energy band diagram* (fig. 2.3-2). The diagram shows the

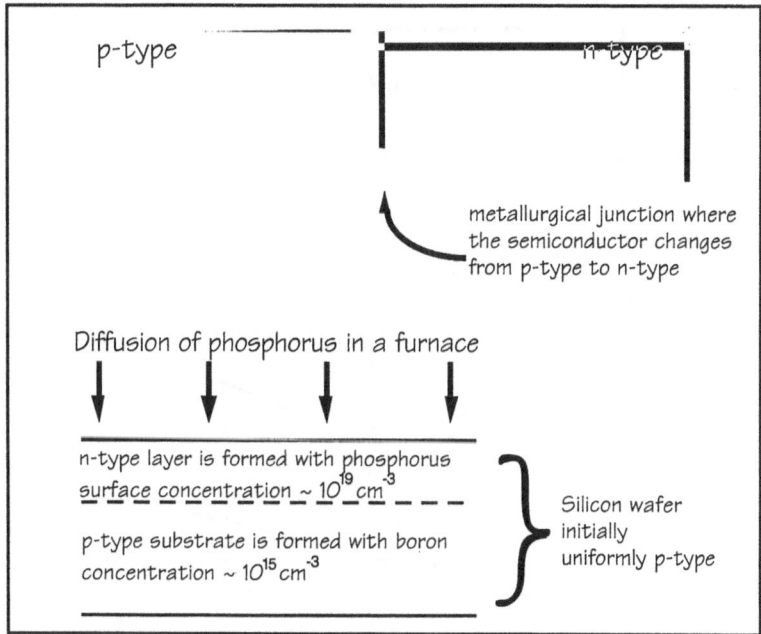

Fig. 2.3-1   A pn-junction is formed by intimate contact between p-
and n-type semiconductor. (a) Schematic of pn-junction structure. (b) A
common method for creating a pn-junction.

energy of electrons and holes as a function of position in a semiconductor sample or
structure. The sample or structure is not necessarily in thermal equilibrium. For thermal equilibrium, uniform doping implies flat bands as in fig. 2.2-3. For non-equilibrium, the bands are not expected to be flat throughout the device. An alternative band
diagram is energy plotted against electron (crystal) momentum (see fig. 2.1-6).

For the pn-junction, the first case to consider is *thermal equilibrium*. By using the
magnetostatic expression for the electric field $\mathcal{E} = -\nabla\phi$ and the definition of the volt as
a Joule per Coulomb, it is seen that the one-dimensional electric field is related to the
band bending by

$$\mathcal{E} = -(1/q)(dE/dx)a_x \qquad (2.3\text{-}1)$$

where E is energy, and a_x is the unit vector. Sudden spatial changes in the band energy
imply regions of strong electric field.

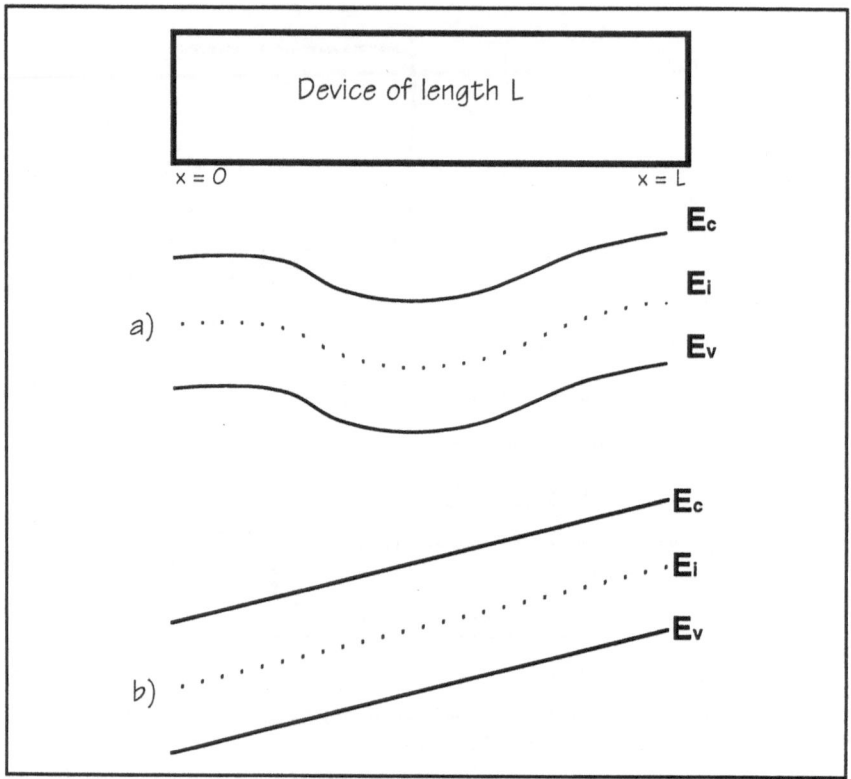

Fig. 2.3-2   Energy band diagram where energy is plotted as a function of position
in the device. (a) Example of a non-uniformly doped device in thermal equilibrium.
(b) Example of uniformly-doped sample under battery bias.

From eq. (2.2-13), a uniformly-doped sample in thermal equilibrium has zero electric
field. Thus, eq. (2.3-1) indicates that a uniformly-doped sample in thermal equilibrium
has flat energy bands, i.e., the bands do not bend. The energies of the electrons at the
bottom of the conduction band and the holes at the top of the valence band are both
constant throughout the sample. However, if the sample is not uniformly doped, as in a
pn-junction structure, there is a *built-in electric field* and the derivative of the energy
with respect to position is non-zero, i.e., the energy bands bend. For thermal equilibrium, the Fermi level is flat across the pn-junction.

The mechanism of the built-in electric field is clearer when one looks at the spatial distribution of the electrons and holes in the vicinity of the plane where the structure changes from n-type to p-type, i.e., the metallurgical junction. When a pn-junction is formed, the concentration gradient of holes on the p-side near the metallurgical junction causes holes to *diffuse* across the junction into the n-side. Immobile negatively-charged boron ions are left behind on the p-side, and are no longer balanced by an equal concentration of holes. Likewise, the concentration gradient of electrons on the n-side near the metallurgical junction causes electrons to diffuse across the junction into the p-side. Immobile positively-charged phosphorus ions are left behind on the n-side, and are no longer balanced by an equal concentration of electrons. Due to the separation of immobile charge, a built-in electric field develops between the exposed positive donor ions and the exposed negative acceptor ions as depicted in fig. 2.3-3(a).

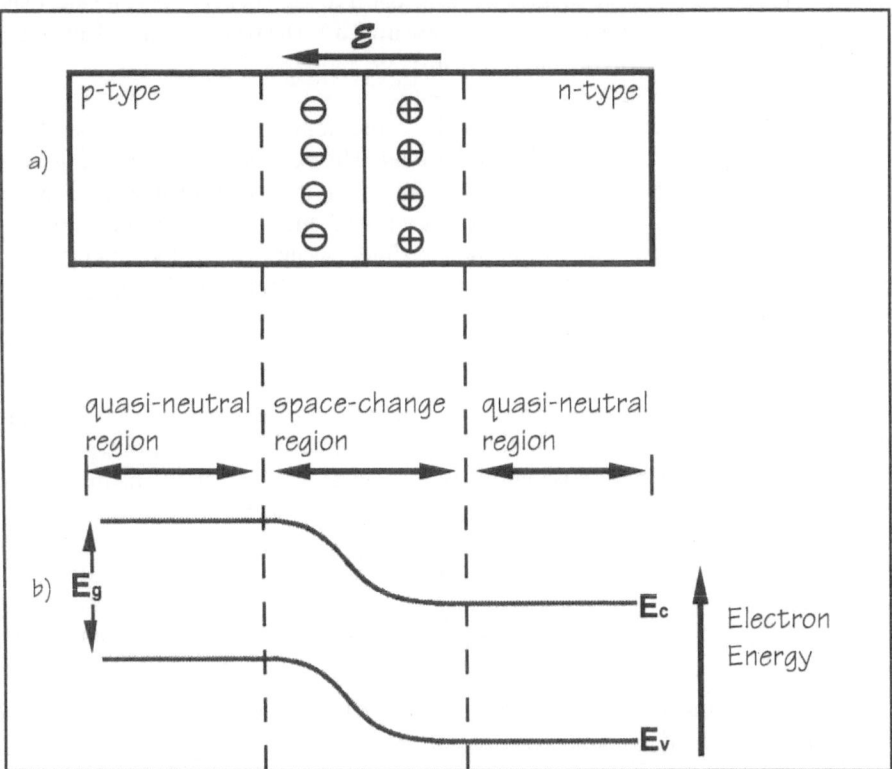

Fig. 2.3-3   The pn-junction in thermal equilibrium, with the individual sides uniformly doped. (a) Built-in electric field extending from the ionized donors to the ionized acceptors. (b) Energy band diagram.

Thus, as the electrons and holes respond to their respective concentration gradients and diffuse across the junction, an internal electric field develops. The field tends to cause carriers to drift in the opposite direction. Eventually, as the junction is formed, the force of this developing built-in electric field becomes *exactly equal and opposite* to the force of the concentration gradients. The diffusion process is self-limiting as the diffusion and drift currents for both holes and electrons balance each other and achieve a thermal equilibrium condition.

The band bending in a pn-junction is caused by the built-in electric field that develops across the junction. Because of the polarity of the field, an electron on the n-side will need to increase in energy if it is to transit the junction and reach the p-side. Likewise, a hole on the p-side will need to increase in energy if it is to get over to the n-side. (The direction of increasing energy for a hole is downward.) The bands bend as depicted in fig. 2.3-3(b), in response to the built-in field.

The region between the exposed donor and acceptor ions is called a *space-charge region* because of the presence of the strong field. This region is also referred to as a "depleted" region because, in thermal equilibrium, all of the carriers have been swept out by the field. (When the junction is forward biased, the space-charge region is certainly not depleted.) Far away from the metallurgical junction, if the doping is approximately uniform, there will be no significant built-in electric field in thermal equilibrium. This is the so-called *quasi-neutral region* of a pn-junction structure, where the bands are flat.

The built-in voltage V_{bi} can be calculated for certain simple doping profiles. For the case of uniform doping on each side of the junction, the electric potential varies strongly in the space-charge region, but is uniform in each quasi-neutral region. The Boltzmann relations, eqs. (2.2-7), at the p-edge of the SCR are then

$$n_{po} = n_i e^{q\phi_{po}/kT} \qquad\qquad (2.3\text{-}2)$$

$$p_{po} = n_i e^{-q\phi_{po}/kT} \qquad\qquad (2.3\text{-}3)$$

where the subscript "po" means p-type material in thermal equilibrium. Likewise, at the n-edge of the SCR

$$n_{no} = n_i e^{q\phi_{no}/kT} \qquad\qquad (2.3\text{-}4)$$

$$p_{no} = n_i e^{-q\phi_{no}/kT} \qquad\qquad (2.3\text{-}5)$$

Division of eq. (2.3-3) by eq. (2.3-5) yields

$$\frac{P_{po}}{P_{no}} = e^{(q/kT)\,(\phi_{no} - \phi_{po})} \qquad (2.3\text{-}6)$$

By using the Law of Mass Action, noting that uniform doping requires $n_{no} = N_{DD}$ and $P_{po} = N_{AA}$, and noting $V_{bi} = \phi_{no} - \phi_{po}$, eq. (2.3-6) can be written as

$$V_{bi} = (kT/q)\,ln[N_{AA}N_{DD}/n_i^2] \qquad (2.3\text{-}7)$$

Figure 2.3-4 shows energy band diagrams for a pn-junction for the three distinct cases of thermal equilibrium, forward bias, and reverse bias. Forward bias means the positive and negative terminals of a battery are connected to the p- and n-sides of the pn-junction, respectively. Reverse bias is the opposite case. During illumination, a solar cell connected to an external circuit behaves as if it were forward biased. Biasing, either by a battery or illumination, produces a dramatic change in the band diagram.

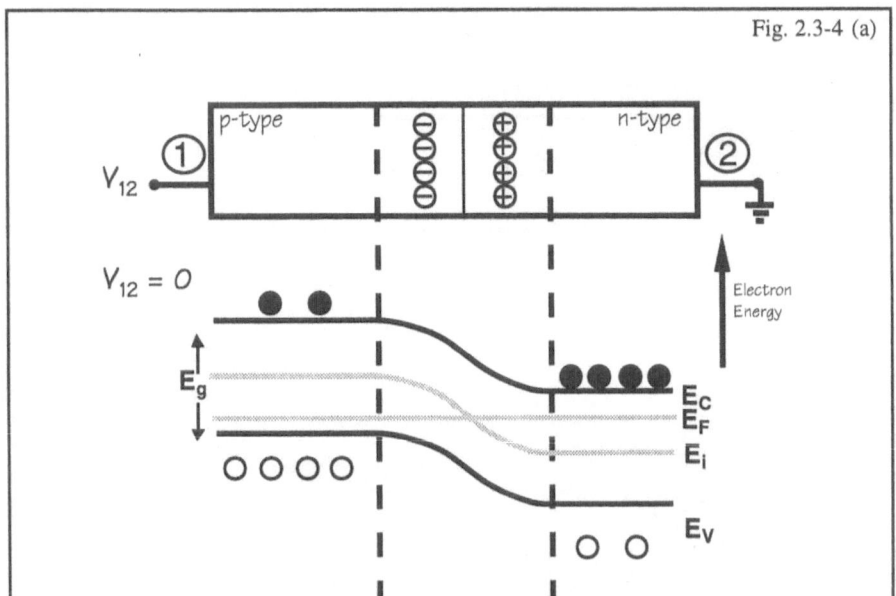

Fig. 2.3-4 (a)

Fig. 2.3-4   The effect of a bias on a pn-junction where each side is uniformly doped. (a) Thermal equilibrium. (b) Forward bias (*next page*). (c) Reverse bias (*next page*).

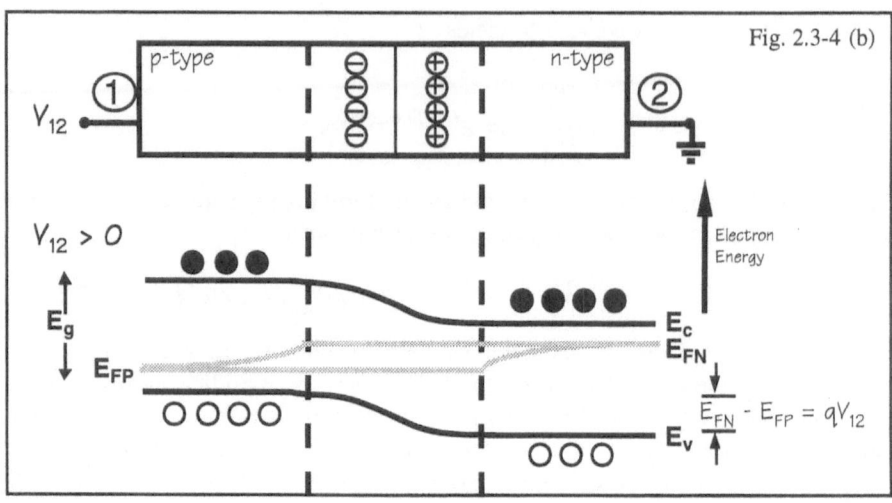

Fig. 2.3-4 (b)

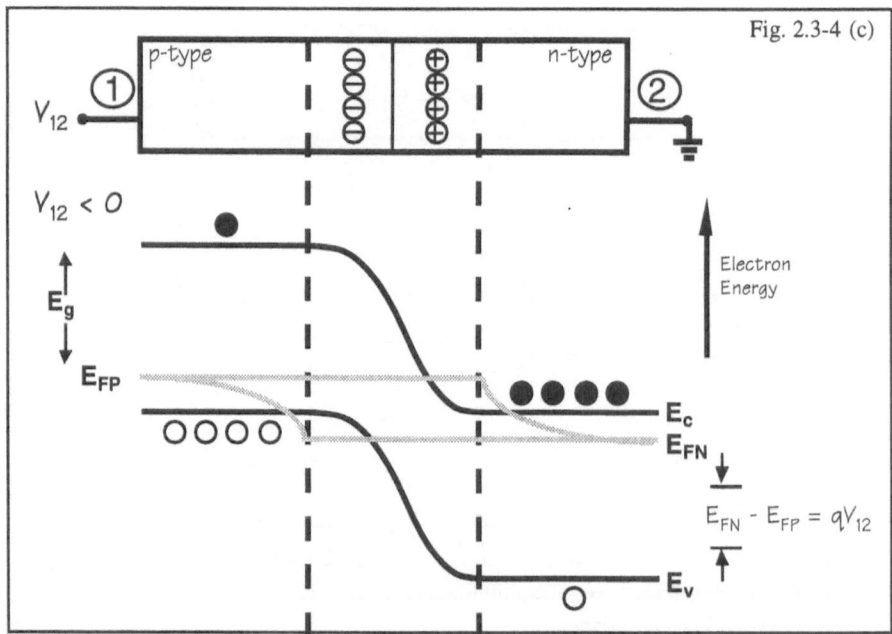

Fig. 2.3-4 (c)

B. Bias Across a PN-Junction Produces a Current

When the junction is *forward-biased*, the Coulomb force (i.e., the electrostatic force) of the charge in the battery tends to push the *majority* carriers away from the terminals of the pn-junction structure. That is, the positive terminal of the battery pushes the holes in the bulk of the p-side toward the metallurgical junction. And the negative terminal of the battery pushes the electrons in the bulk of the n-side toward the metallurgical junction. This has the effect of "covering up" some of the exposed donor and acceptor ions at the edges of the space-charge region. Some of the immobile ions are now neutralized by the proximity of carriers. Consequently, the strength of the built-in field across the junction, and the gradient of the band bending, decreases. As shown in fig. 2.3-4(b), this causes the net band bending to decrease. Built-in field across the space-charge region is partially defeated. Diffusion of majority carriers across the junction now exceeds the tendency of those carriers to drift back in the opposite direction. This can be correlated to changes in the Fermi level. For the non-equilibrium case of forward bias, the Fermi level splits into *quasi Fermi* levels because the electrons and holes are not allowed to relax and come into thermodynamic equilibrium with each other. With the assumption that all of the applied bias is dropped across the space-charge region, the difference in the quasi Fermi levels for the quasi-neutral regions will equal the bias across the diode: $E_{FN} - E_{FP} = qV_{12}$. Alternatively, $V_{12} = V_P - V_N$, the difference of the quasi Fermi potentials. The result is that majority carriers flood across the junction, and constitute the dc steady-state current seen in a forward-biased pn-diode. See fig. 2.3-5(a).

If the junction is *reverse-biased*, just the opposite happens. The Coulomb force of the battery tends to attract *majority* carriers away from the edges of the space-charge region. This causes more donor and acceptor ions to become exposed. The strength of the built-in field across the junction increases with a subsequent increase in the band bending gradient. The net band bending increases. This is shown in fig. 2.3-4(c). The diffusion of majority carriers across the junction is now exceeded by the tendency of those carriers to drift back in the opposite direction. As a consequence, there is essentially no majority carrier current across the junction. The only current comes from the few minority carriers (supplied at the battery terminals) that are able to reach the vicinity of the pn-junction and "fall down" the energy barrier. Again, the Fermi level is replaced by quasi Fermi levels, and the bias across the diode is equal to the difference in the levels: $E_{FN} - E_{FP} = qV_{12}$, i.e., $V_{12} = V_P - V_N$. Total current in the circuit is very small, and, at least for a silicon junction, almost independent of the magnitude of the reverse bias. That is, the current "saturates" in the reverse direction. See fig. 2.3-5(b). Accordingly, the reverse-bias diode current is called the pn-diode *saturation current*, I_o.

For the non-equilibrium situation of forward or reverse bias, the difference in the quasi Fermi potentials is precisely the quantity measured by a voltmeter across the diode. In thermal equilibrium, the quasi Fermi levels reduce to the Fermi level. The Fermi level must be constant across the diode in thermal equilibrium. Thus, in thermal equilibrium, a voltmeter placed across the diode would measure *zero voltage*. The built-in voltage across the diode *can not* be measured by a voltmeter. Kirchoff's voltage law demanding that the sum of the voltages around a closed circuit is zero is satisfied for the diode in thermal equilibrium by noting that the built-in voltage (electrostatic potential) exactly offsets the chemical potential across the junction.

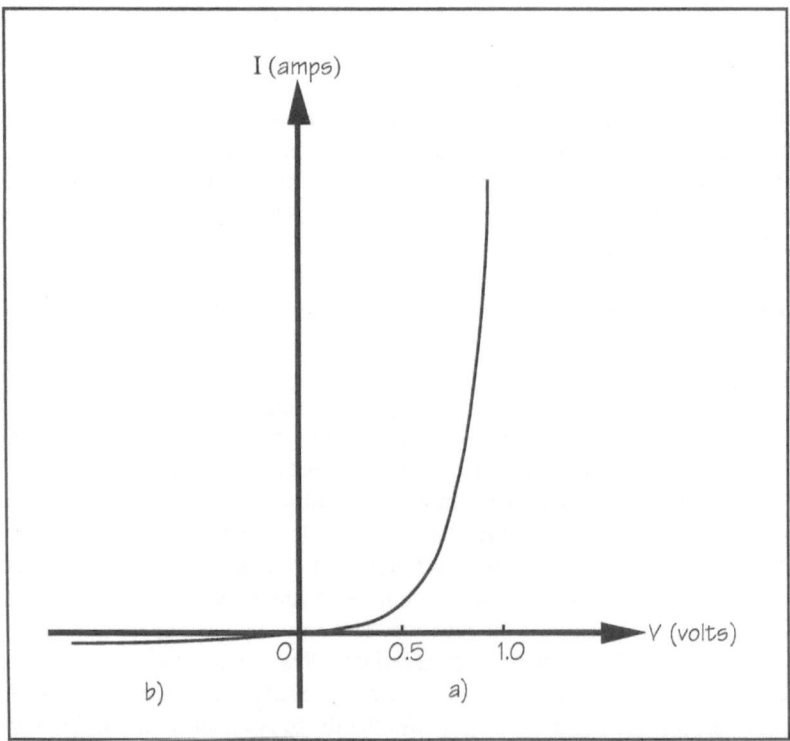

Fig. 2.3-5   Typical current-voltage (I-V) curve for a silicon pn-junction diode. The current is proportional to the area of the diode. (a) Forward bias. (b) Reverse bias.

C. Majority and Minority Carriers Move by Different Mechanisms

Majority carriers move mostly by drift, i.e., Coulomb or electrostatic force. This has an interesting consequence for the way excess majority carriers flow out of the contacts of the diode and into the external circuit. The Maxwell equation relating electric flux density $\mathbf{D} = (D_x, D_y, D_z)$ and free charge ρ is given by div $\mathbf{D} = \rho$. Here div is the divergence operator; div $\mathbf{D} = (\partial/\partial x)D_x + (\partial/\partial y)D_y + (\partial/\partial z)D_z$. This indicates the net outward flux of $\mathbf{D}$ per unit volume at a point is equal to the charge density at that point. Since $\mathbf{D}$ is related to electric field strength $\mathcal{E}$ through the constitutive relation $\mathbf{D} = \epsilon \mathcal{E}$, where ϵ is the electric permittivity of the material, this becomes in one-dimension $d\mathcal{E}_x/dx = \rho/\epsilon$. For any excess majority charge distribution in the quasi-neutal regions, there will be a corresponding spatial variation in electric field. Such local non-uniformities can be caused by doping gradients in the quasi-neutral regions, and are not necessarily an indication of a non-thermal equilibrium condition. However, if excess majority charge density comes about in the QNRs by an external event, the diode is no longer in thermal equilibrium. The diode will respond to the sudden appearance of excess charge in an attempt to smooth out the associated spatial variations in electric field and regain thermal equilibrium. If a transient event produces excess majority carriers in the QNR, the excess majority carriers quickly push each other away so that the charge density variation decays. A transient current would be seen in the external circuit. When the excess charge decays to zero, the associated transient field decays to a constant (which can be taken as zero). For a steady-state event that produces excess majority carriers, the diode is not allowed to return to thermal equilibrium, and the Coulomb force between majority carriers creates a steady-state current in the circuit. The tendency to smooth out non-uniform electric-field variations in the QNRs, i.e., excess majority charge density, occurs very quickly and is a consequence of the *continuity equation* for charge. Magnetic field strength $\mathbf{H}$, current density $\mathbf{J}$, and electric flux density $\mathbf{D}$, are related by the Maxwell equation

$$\nabla \times \mathbf{H} = \mathbf{J} + d\mathbf{D}/dt$$

By taking the divergence of both sides and noting that the divergence of the curl is identically zero, this yields

$$0 = div\ \mathbf{J} + \frac{(d)}{dt} div\ \mathbf{D}$$

Substitution of div $\mathbf{D} = \rho$ yields the continuity equation for charge in the *absence* of generation and recombination:

$$0 = div\ \mathbf{J} + d\rho/dt$$

This says that the net outward flux of charge at a point is equal to the rate of appearance of charge at that point, i.e., conservation of charge. Substitution of the constitutive relations $\mathbf{J} = \sigma\mathcal{E}$ where σ is conductivity and $\mathbf{D} = \epsilon\mathcal{E}$, and the relation div $\mathbf{D} = \rho$, yields

$$0 = d\rho/dt + \sigma\rho/\epsilon$$

The solution is

$$\rho(t,x) = \rho(0,x)e^{-t/\tau} \tag{2.3-8}$$

where $\tau = \epsilon/\sigma$ is the *dielectric relaxation time* and characterizes the rate at which excess majority carrier charge density tends to decay. For silicon, $\epsilon = \epsilon_{si}\epsilon_o$, where $\epsilon_o = 8.85 \times 10^{-14}$ farads/cm is the electric permittivity of free space and ϵ_{si} is the relative permittivity of silicon. As an example, if the p-type QNR of a diode is doped 1.5×10^{16} cm^{-3}, then the resistivity is 1 Ω cm, the conductivity σ is 1 mho/cm, and the value of τ is about 1 ps. Thus, the diode instantly responds to excess majority charges by forcing them out through the contacts through a reconfiguration of the charge density.

Minority carriers in the QNRs, on the other hand, move mostly by diffusion. This results from the shielding effect of majority carriers. When a *minority* carrier appears, it attracts a huge number of majority carriers around it. They effectively screen much of an electric field that might be present. Consequently, minority carriers in the quasi-neutral regions of a structure often do not "feel" an electric field. Instead, they move mostly in response to concentration gradients. This is consistent with the small dielectric relaxation time. The disturbance to space-charge neutrality caused by the injection of excess minority carriers is immediately cancelled by a reconfiguration of the majority carriers.

D. Recombination in Silicon

Excess carriers are created in pairs: $\Delta n = \Delta p$. Thus, the total concentrations of electrons and holes are $n = n_o + \Delta n$, and $p = p_o + \Delta p$, respectively, where the subscript denotes equilibrium values. There are two cases, low injection and high injection. *Low injec-*

tion means there is a well-defined minority carrier, i.e., $\Delta n = \Delta p$ is much less than the majority carrier concentration. For low injection, the excess carriers will have little effect on the majority carrier concentration. However, because the thermal equilibrium concentration of minority carriers is often eight or more orders of magnitude less than the majority concentration, the minority concentration can be drastically affected by creation of excess carriers. As an example, consider the p-type QNR of a diode initially in thermal equilibrium. Assume the p-QNR is uniformly doped 10^{16} cm^{-3}, has a volume of 0.01 cm^{-3}, and is illuminated for 1 msec with a 1 mW green laser ($\lambda = 0.55$ μm). The minority electron density before the pulse is 2.3×10^{4} cm^{-3}. The total energy deposited in the p-QNR is 10^{-6} J with each photon having 2.26 eV energy. This produces 2.8×10^{12} pairs in a volume of 0.01 cm^{-3}, which is 2.8×10^{14} pairs cm^{-3}. The minority concentration has increased ten orders of magnitude, but the majority concentration is negligibly affected. When an *excess* minority carrier appears, there is a certain probability it will recombine with a majority carrier. For low injection, the rate of recombination is dependent on the availability of minority carriers. In silicon, because recombination occurs through intermediate levels in the bandgap, the recombination rate R of excess carriers is expected to be proportional to the excess minority carrier concentration Δn, the concentration of intermediate states (traps) in the bandgap N_t, and the speed imparted to carriers by the temperature of the lattice, or thermal velocity v_{th}. If electrons are the minority carrier, this is

$$R = \Delta n N_t v_{th} \sigma_n \ ,$$

where σ_n is the proportionality constant.

Here, as a first approximation, the intermediate states are assumed to be near the center of the bandgap where they are most effective. The thermal velocity is the average speed of electrons as they randomly collide with lattice atoms, impurity atoms, and other defects. Since this is a random walk, there is no net movement over a long period of time. From statistical mechanics, the kinetic energy in thermal equilibrium is $3kT/2$. Kinetic energy also equals $mv_{th}^2/2$, where m is the effective mass of an electron. This implies $v_{th} = (3kT/m)^{1/2}$. For silicon at 300 K, $v_{th} = 10^7$ cm/s. The proportionality constant σ_n has the units cm^2 and represents the effective cross section for the capture of a minority carrier. If the minority carrier wanders within the area σ_n about the intermediate state, it recombines. This is expected to be on the order of atomic dimensions, roughly $(5 \times 10^{-8}$ cm$)^2$, i.e., the order of 10^{-15} cm^2. The product $v_{th}\sigma_n$ has the units cm^3/s and represents the volume displaced by a minority electron per unit time if the cross section of the volume has the area σ_n. If the recombination site resides in that volume, a minority carrier would be captured per second. R has the units of cm^{-3}s^{-1} and can be expressed as

$$R = \Delta n / \tau_n$$

where $\tau_n = N_t v_{th} \sigma_n$ is an effective minority carrier *lifetime*. This is the median time to recombination for a minority carrier. In p-type float-zone silicon doped 10^{16} cm^{-3}, τ_n is on the order of 300 μs. A related parameter is the minority carrier diffusion length $L_n = (D_n \tau_n)^{1/2}$, where D_n is electron diffusivity. This is the one-dimensional mean length to recombination for a minority carrier. For *high injection*, $\Delta n = \Delta p \gg n$ and p, and there is no minority carrier. The recombination rate then becomes $R = \Delta n / (\tau_n + \tau_p)$. Under one-sun illumination, solar cells operate in low injection. Ten to 100 suns may be required to reach the high injection condition.

Recombination also occurs at the surface of a diode where the lattice abruptly terminates. The discontinuity at the surface introduces a large concentration of recombination sites. For low-injection conditions, minority carriers flow into the surface with a speed given by $S = v_{th} \sigma N_t$, where N_t is now the areal density of recombination states at the surface. The corresponding minority carrier current density flowing into the surface is $J = qS\Delta n$, where Δn is the minority concentration at the surface. Surface recombination velocity is limited to the thermal velocity. Large values of S mean that excess carrier densities at the surface are immediately removed through recombination.

In general, for solar cells, it is advantageous to minimize all recombination rates. In a solar cell, any recombination degrades the collection of photogenerated minority carriers and subtracts from the output current. Bulk recombination is sensitive to dislocations in the lattice and to impurities that introduce deep levels in the bandgap. High-temperature treatments severely degrade lifetime by enhancing the movement of impurities through the bulk. This creates mechanical stress and subsequent dislocations. Quenching after heat treatment makes this worse. This produces a problem for the diffusion of intentional impurities to form a pn-junction. The necessary slow cool-down makes the desired formation of a shallow diffusion difficult. Surface recombination is sensitive to surface treatments and films. Etchants that produce smooth surfaces and oxides that tie up dangling bonds can lower S values by several orders of magnitude. Values in the range of 100 cm/s have been obtained for the top surfaces of solar cells. Additionally, hydrogen treatments passivate both surface and bulk recombination states. This is achieved either by annealing in forming gas (10% H_2, 90% N_2) or by deposition of silicon nitride from a silane and ammonia vapor. The pn-junction in a solar cell plays the critical role of separating the excess *minority* electrons and holes that are generated in the p-side and n-side quasi-neutral regions, respectively, when the device is exposed to sunlight. In order to be separated and collected as useful current, the minority carriers in a quasi-neutral region must survive long enough to diffuse over to the pn-junction. Consequently, the efficiency of a solar cell is highly dependent on the effective minority carrier lifetime. In general, for any semiconductor structure, the *effective* minority carrier lifetime is maximized by maximizing the bulk lifetime and minimizing the various surface recombination velocities.

2.4 STARTING MATERIAL FOR SOLAR CELLS

Commercial silicon solar cells are made from four generic types of material: single-crystal ingots, cast polycrystalline ingots, polycrystalline ribbons, and polycrystalline recrystallized thick films on inexpensive substrates. Recent developments include ribbons that are almost single crystal in structure. These are all crystalline materials and are grown by a variety of processes. Single-crystal ingot material, i.e., single crystal material sliced from ingots yields the highest efficiency cells, but is also the most expensive choice. High efficiency is afforded by the inherently low density of crystal defects and unintentional impurities coupled with a sufficient thickness (several hundred microns) for absorbing much of the usable light. However, growth of ingot-based single-crystal material is energy intensive and requires costly surface preparation before cell processing begins. Cast poycrystalline silicon is also an ingot-based material, but the ingots are cheaper to produce – with the drawback of slightly lower efficiency. Polycrystalline ribbons and recrystallized thick films are not ingot-based. They avoid the cost of sawing and extensive surface preparation, but they have the drawback of low fabrication throughput rates.

Starting material for all cell processing is *electronic-grade* silicon. It has a total impurity concentration on the order of 10 parts per million atoms. Electronic-grade silicon is made from *metallurgical-grade* silicon, which has a total impurity concentration between 10 and 20 parts per thousand atoms. These terms are somewhat arbitrary and denote the stage of purification more than the exact impurity concentration in the material. There are several stages of refinement and crystal growth.

Metallurgical-grade silicon is made by the carbothermic reduction of silicon dioxide in an electric arc furnace process. A crucible is filled with quartz sand and high-carbon matter such as coal. The mixture is electrically heated to liquification by an electrode that remains submerged below the surface of the melt. Numerous reactions take place, with the overall reaction being

$$2 \text{ C (solid)} + \text{SiO}_2 \text{ (solid)} \longrightarrow \text{Si (liquid)} + 2 \text{ CO (gas)}.$$

Liquid silicon sinks to the bottom of the crucible and is drawn off and allowed to solidify. Major impurities for metallurigical grade silicon are boron (15-25 ppmw), phosphorus (30 ppmw), and iron (0.5 -0.7 wt%)[13].

Metallurgical-grade silicon is ground into a powder, and reacted with anhydrous HCl gas in a fluidized bed reactor to produce trichlorosilane ($SiHCl_3$) by the reaction

$$Si \text{ (solid)} + 3 \text{ HCl (gas)} \xrightarrow{250 C} SiHCl_3 \text{ (gas)} + H_2 \text{ (gas)}.$$

Trichlorosilane liquifies at 31.8° C and can be highly refined from various impurity chlorides by successive fractional distillation. This yields impurity concentrations, particularly those from groups III and V of the periodic table, below 1 ppb. The ultra-pure trichlorosilane liquid is vaporized and introduced into a quartz bell jar along with hydrogen. The bell jar holds a thin electronic-grade polycrystalline silicon rod in an inverted-U configuration. Electrodes are connected to the polysilicon rod and it is resistively heated to between 1000 and 1100° C. Subsequent hydrogen reduction of trichlorosilane yields electronic-grade silicon by chemical vapor deposition onto the heated polycrystalline silicon U-shaped substrate:

$$2 \text{ } SiHCl_3 \text{ (gas)} + 3 \text{ } H_2 \text{ (gas)} \xrightarrow{1000 C} 2 \text{ Si (solid)} + 6 \text{ HCl (gas)}.$$

This approach was developed by Siemens in the mid-1950s[14], and is still the dominant process for producing electronic-grade silicon. The polysilicon rod has an initial diameter of about 4 mm. When the rod reaches about 16 mm in diameter, the process is terminated, and the rod is ready for use as feedstock for single crystal ingot growth. The advantage of solidification around a rod, as opposed to a crucible, is that there is no contact contamination. Additional polysilicon rods, 4-6 mm in diameter, for future depositions (and for seeds to initiate single-crystal growth) can be drawn from the hot cores of finished rods[15] (slim-rod pulling).

Electronic-grade silicon is priced at \$40-\$60/kg (1995). Because solar cells are large area devices, cell manufacturers can not afford to grow single-crystal ingots or cast polycrystalline ingots directly from virgin electronic-grade polysilicon feedstock. Instead, they use the discarded ingots, parts of ingots, and residue left over from ingot growth (collectively called "scrap") that are generated by the much larger silicon integrated circuits industry as feedstock for their own ingot or ribbon growth. Scrap material consists mostly of the tops and tails of Czochralski ingots (described below) and residue from growth crucibles (pot scrap). Additionally, a major supply of wafers for single-crystal cells comes from the purchase of 6-inch and 8-inch control wafers that are of no further use to the silicon IC industry. During IC fabrication, many control wafers are used to characterize the progression of the integrated circuit process. In IC manufacturing, only the top three or four microns of the wafer are used for device fabrication. The rest of the wafer serves as mechanical support for this active layer. After parametric measurements are made to characterize a particular step in the processing, the control wafer is sent to a facility for reconditioning into a new control wafer. This is done by removing the top few microns of the wafer with either chemical etching or mechanical abrasion, followed by removal of surface damage and polishing. The control wafer can be recycled several times. Eventually, after a certain number of cycles, the wafer becomes too thin for further convenient use. At this point, the wafer is either recycled into a Czochralski melt (if process-added impurities have not made this unsuitable), discarded, or sold to a photovoltaic manufacturer. The PV producer then removes the processed top layer of the wafer by sandblasting. This is followed by a NaOH etch to remove the surface damage and render a planar specular surface. As IC wafer diameters become larger (8-inch is now common), it becomes increasingly worthwhile for an IC manufacturer to reclaim control wafers to the maximum extent possible. Thus, as the IC industry moves toward 10-inch wafers, the cost of reject control wafers is expected to increase. All silicon feedstock for production of crystalline silicon cells comes from these sources. By 1995, top and tails of ingots were becoming very scarce, and pot scrap was becoming the primary source of silicon feedstock for growth of PV Czochralski ingots. As of 1996, there were *no* wafer suppliers in the U.S. that grew crystalline silicon material strictly for the solar cell industry.

In 1995, pot scrap material costs were between *$10 and $15 per kg*, and amounted to *about 1250 metric tons*[16]. Silicon requirements for ingot-based cell manufacture are about *20 kg of starting material per kW* of module output power. Thus, the 1995 capacity of low-cost silicon for solar cell manufacture was approximately 63 MW per year. The rest (of the total 81 MW in 1995) presumably came from reject control wafers and a few tops and tails. Robust material supply is a *major advantage* that the silicon solar cell industry has with respect to any competitive PV technology: the material is already available from a sister industry. In this way, silicon solar cell manufacturers are the direct beneficiaries of a much larger and established industry. However, demand is increasing faster than capacity, and a shortage of feedstock for crystalline silicon cells is expected by the year 1998. The problem is somewhat ameliorated by the trend toward thinner (200 μm) cells that make extensive use of surface texturing for light trapping. When demand does exceed capacity, cell manufacturers will either have to buy expensive virgin electronic-grade feedstock or depend on silicon suppliers to develop a less expensive so-called "solar grade" material. For the former material, the average cost of about $50/kg corresponds to about $1/W in the finished cell. This is a prohibitively expensive option for an industry that requires lower production costs for industry expansion.

Solar grade is undefined, but has a purity and defect density slightly less than electronic grade. Estimates are that such material would be commercially viable when the lack of scrap pushes the price to about $20/kg and guarantees a market of several thousand metric tons per year. The minimum technical requirement for solar grade is the ability to achieve 1 Ω cm resistivity material. This restricts boron content to about 0.3 ppm, phosphorus content to about 0.1 ppm, and carbon to about 4.3 ppm[17]. Overall purity is not much worse than 10 parts per million atoms. Polycrystalline cells with rather low, but commercially acceptable, efficiency of 11% have been made from material that slightly exceeds these requirements. Because of the proximity of graphite fixtures in the common ingot growth process, carbon concentration is a major impurity. Studies have shown that solar-grade wafers yield comparable performance to electronic-grade wafers if the carbon concentration is kept below 25-30 ppmw[18].

A. Single-Crystal Silicon - Efficient and Expensive

Single-crystal ingots are grown by the Czochralski or float-zone processes and are cylindrical in shape. These processes yield round wafers with the ingot sliced perpendicular to its axis. Round wafers afford a lower packing density in a module compared to square wafers. To maximize packing density, ingots grown for solar cell purposes are milled into a square cross section before wafer slicing. The excess silicon can be reused for future ingot growths.

Some large producers of single-crystal cells have extensive crystal growth operations strictly for their own fabrication line. This requires a major capital investment. A number of smaller fabricators of single-crystal cells simply purchase wafers. Purchase of partially processed reject wafers from an IC fab line is appealing due to the low cost. These wafers are up to ten inches in diameter and 300 to 400 microns thick. (As wafers become larger, the thickness is also increased to afford sufficient mechanical strength. Otherwise, there would be a high rate of wafer breakage during processing.) They have undergone some processing in a facility that might be making memory or microprocessor chips. Integrated circuit fabricators make parametric measurements on their wafers as they go through the processing sequence so they can quickly spot a processing step that has failed to meet specification. If a step has failed, the wafer is "re-worked", i.e., the step is repeated. When this is not possible, the whole wafer is discarded. A small recouperation of wafer costs is made by selling the unusable wafer to the solar cell industry. As with reject control wafers, the cell manufacturer abrades the top layer by sandblasting and saws the wafer to make it square. Etching in sodium hydroxide solution (300 g/l at 80°C) removes the twenty or thirty micron-thick damaged surface. Because the material properties of reject wafers vary from lot to lot depending on the microelectronic circuits that were originally intended for those wafers, a variation in the efficiencies of the subsequent cells is expected. This does not represent a major drawback. In fact, there is always a variation in cell efficiencies regardless of lot-to-lot uniformity of the starting wafers. The final solar cells are binned according to current output before being assembled into photovoltaic modules. By doing this, all the cells in a module are roughly the same – an important requirement for an efficient module.

Whether produced in-house by the cell manufacturer or purchased as reject material from an IC manufacturer, commercial single-crystal silicon is always made by one of two energy-intensive processes: *Czochralski* growth and *float-zone* growth. Czochralski growth is the cheaper and more common of the two; float-zone growth yields the purest and subsequently highest grade silicon material. As of 1996, the most efficient silicon cells and modules were made from float-zone material. These cells and modules achieve efficiencies under 1-sun AM0 illumination of 24% and 20%, respectively. Production-line cells and modules display values closer to 14% and 13%, respectively. In 1995, single-crystal silicon was still the most common material for photovoltaic modules, accounting for about *57% of the year's total worldwide production* of 81 MW.

In the *Czochralski* single-crystal ingot technique (fig. 2.4-1a), chunks of electronic-grade polycrystalline silicon are loaded into a crucible made of vitreous silica (fused quartz)[19], a form of SiO_2. The crucible is fabricated by arc-fusion spinning. Fused quartz is the only good choice of materials due to the reactiveness of molten silicon. Molten silicon is a universal solvent. Crucibles are as large as 300-mm in diameter and have charges as large as 150 kg[20]. The crucible sits on a graphite pedestal that is rotated and also moved along the z-axis. The sides of the crucible are supported by a graphite holder. High purity graphite is used to minimize out-gassing of contaminates during the growth process. The entire assembly – pedestal, holder, and crucible – sits inside a partially evacuated chamber with about 10-20 Torr argon ambient. A high power resistance heater (100 to 350 kW), surrounds the graphite holder. It is important for the quartz crucible to be carefully supported by the graphite holder because quartz becomes soft well below the silicon melting point.

After the system is sealed and put under an argon ambient, power is applied and the

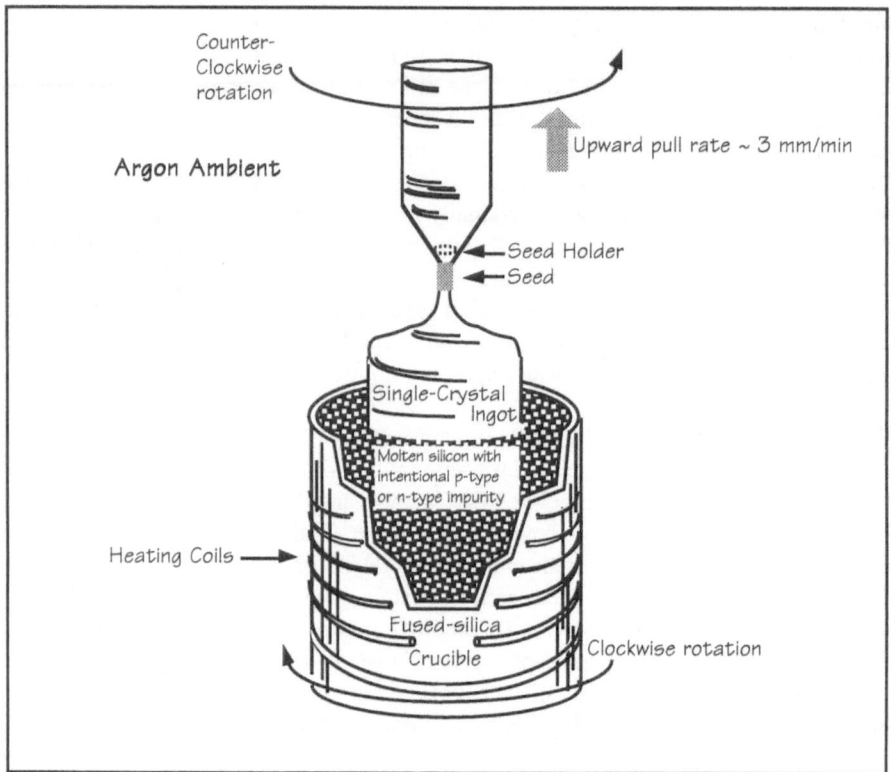

Fig. 2.4-1 (a)   Bulk single-crystal silicon is most often grown by the Czochralski technique.

polycrystalline chunks are brought to a temperature just above the silicon melting point (1415° C). During the growth process, any *intentional* impurities that might be added to adjust the overall electrical resistivity of the ingot, are added to the melt. To initiate ingot growth, a seed crystal with the desired crystallographic orientation, about 6 to 12 mm in diameter, is positioned at the meniscus of the melt. The <111> or <100> crystallographic orientations are chosen because these allow dislocation-free growth. A molybedenum fixture attached to a stainless steel shaft holds the seed. A motor rotates the shaft and the assembly can be moved upward along the z-axis. The seed is then dipped into the melt and thoroughly wetted. With the crucible pedestal rotating in one direction, and the shaft rotating in the opposite direction, the crystal is slowly pulled upward. A large single crystal of silicon, with the same crystallographic orientation as

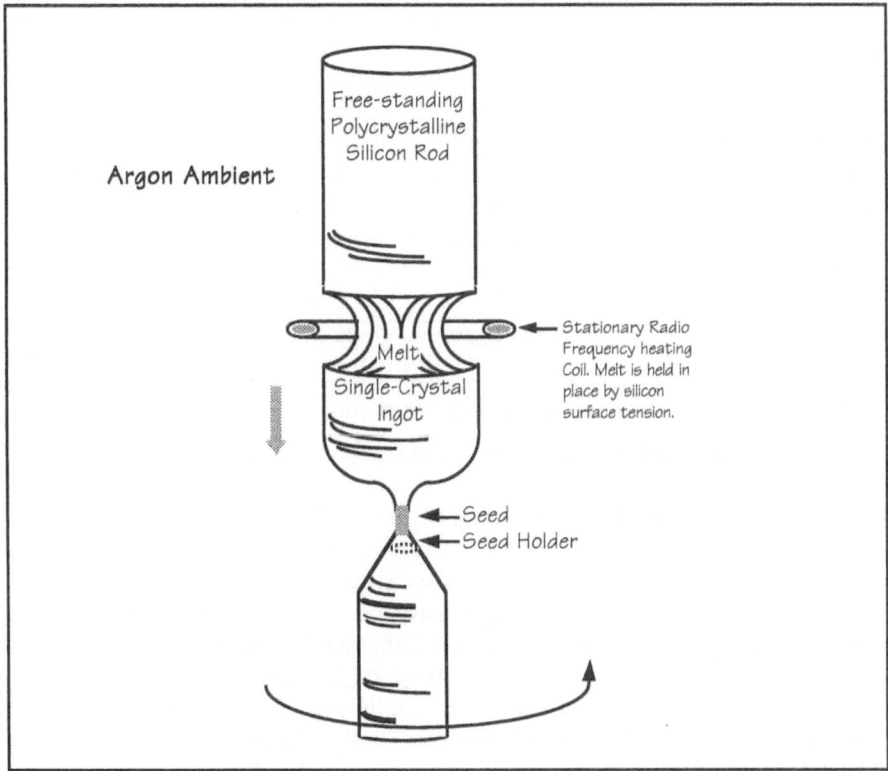

Fig. 2.4-1 (b)   An alternative method is float-zone refining.

the seed, forms at the liquid-solid interface. Initial growth uses a fast pull rate (30 cm/hr) to create a neck about 3 mm wide and 25 mm long. This procedure is necessary to avoid dislocations in the seed crystal from propagating into the ingot. After the neck is created, the pull rate is slowed to about 2.5 cm/hr to achieve the desired ingot diameter. As the melt surface moves downward with growth, the pedestal is elevated so that the growth region is kept in the same thermal configuration with respect to the fixed-position resistance heater. The vertical pull rate, rotation rate, and temperature gradients in the developing crystal determine the diameter and quality of the ingot. Growth is terminated before the charge is completely depleted to avoid thermal shock and subsequent deformation of the crystal. To minimize this effect, the end of the ingot (tail) is tapered similar to the neck. After the ingot is pulled free from the melt, it slowly cools to room temperature. An ingot 15 cm in diameter is pulled at an average of about

7.0 cm/hr. This corresponds to a mass solidification rate of 2.9 kg/hr, which is somewhat slower than that for cast polycrystalline ingots, but is about an order of magnitude faster than the solidification rate for silicon ribbons. Fully-grown ingots 1 meter in length, 25 cm in diameter, and 70 kg in mass are easily achieved. The ingot is one large single crystal of silicon, with an overall purity level in the several to 10 atoms per million range. Crucibles can only be used once because they crack during cool-down. This is related to mechanical mismatch during solidification. Additionally, molten silicon slowly dissolves the crucible. Crucible replacement and electric power are the major expenses for the Czochralski process.

A Czochralski ingot differs from the electronic-grade silicon from which it is made in *two fundamental ways*:

    (i) it is single-crystal material, as opposed to polycrystalline material;

    (ii) the ingot has a lower concentration of unintentional metal impurities.

Total impurity concentration of the ingot, excluding carbon and oxygen, is about the same as for electronic-grade feedstock – several parts per billion atoms[21]. Oxygen is introduced by the slow dissolution of the fused silica crucible from which the ingot is grown. It remains in the crystal at a concentration of a few parts per million atoms. Carbon is introduced by the graphite heater and other graphite fixtures in the growth chamber. It appears in the ingot at less than 0.5 parts per million atoms. Thus, while the concentration of most impurities will decrease as the ingot is grown, the concentrations of oxygen and carbon will increase. This is not totally deletereous. Unintentional oxygen can serve as a "gettering" mechanism to attract various other impurities and crystal defects during high-temperature processes.

The great improvement in silicon purity with respect to metal contaminants is attributable to the *impurity distribution coefficient* k_o (sometimes called the segregation coefficient):

$$k_o = C_S/C_L$$

where C_S is the equilibrium solubility of the impurity in the silicon solid, and C_L is the equilibrium solubility of the impurity in the silicon liquid (melt). For most elements, the distribution coefficient at the solid/liquid silicon interface is less than 1. During the growth of the ingot, these impuriites are segregated into the melt. As the ingot grows, the melt gets progressively richer in those impurities for which $k_o < 1$. When the growth process is finished, most of the impurities are either left in the remaining liquid or reside in the end (tail) of the ingot. The end of the ingot is sawed off and discarded. For the elements boron, phosphorus, and arsenic, though, the impurity distribution coefficients are relatively close to unity (0.8, 0.35, and 0.30, respectively). This means that if any of these impurities are intentionally added to the melt to adjust the electrical resistivity, the resulting concentration of the impurity in the ingot will be approximately uniform.

Float-zone growth[22] was developed in the early 1950s at Siemens and other laboratories. This is a crucible-free method, in which a liquid zone is suspended between freezing and melting interfaces in a slightly positive-pressure argon ambient (fig. 2.4-1b). A rod of electronic-grade polycrystalline silicon is held upright in an argon ambient. The rod is in contact with a seed crystal at its bottom end and the entire assembly is moved vertically downward through a fixed-position radio-frequency inductive heating coil operating near 4 MHz. During the growth process, the rod and seed are rotated in opposite directions, several revolutions per minute. As the rod passes downward through the heating coil, the silicon melts in the immediate vicinity of the coil, with solid above and below the melt zone. The liquid silicon at the solid/liquid interface is 1-3 cm thick and kept just a few degrees above the melting point. As the melt passes below the ring and cools, it solidifies into single crystal, with the same crystallographic orientation as the seed. Similar to Czochralski growth, impurities are segregated into the liquid fraction. The liquid fraction eventually reaches the top end of the rod leaving behind an ingot of very pure single crystal, free of dislocations[23]. There are several doping methods for adjusting the resistivity. Core doping is commonly used to dope the ingot with boron. In this method, the polycrystalline rod has been deposited around a boron-doped core during the Siemens inverted-U process. (Cores come from initial rods that were doped by adding BCl_3 gas to the $SiHCl_3$-H_2 mixture in the Siemens process.) During float-zone growth, convection currents in the rod produce uniform mixing of the boron with the undoped part of the rod and a subsequent uniform boron concentration in the ingot.

Two forces prevent the melt from spilling out of the polycrystalline rod as it moves through the coil. These are the surface tension of the molten silicon, and the electrodynamic pressure caused by the time-varying magnetic field from the coil. At the high temperatures near the coil, silicon atoms are strongly ionized. The circularly moving positively-charged silicon ions feel a force v x $\mathbf{B}$ from the field which limits their outward radial movement. For smaller diameter ingots, surface tension plays more of a role in containing the liquid. However, the forces of gravity and the angular acceleration of the rod tend to destabilize the melt. This imposes a limiting value on the diameter of the ingot. Meter-long ingots in excess of 150-mm diameter have been successfully grown.

The main advantage of float-zone material over Czochralski is that *no crucible* is needed to contain the melt. Float-zone ingots avoid oxygen contamination that comes from

quartz crucible contact and carbon contamination associated with graphite holder proximity. Minority carrier lifetime is highly sensitive to the growth and cooling rates of the
ingot. A large ratio of growth rate to thermal gradient allows swirl-free growth, while a
small product of these parameters minimizes defects caused by cooling. By optimizing
growth conditions, lifetimes higher than 10 ms have been measured[24]. In low-doped p-
type silicon, this corresponds to a diffusion length of several thousand microns.

After the Czochralski or float-zone ingot is grown, the seed end and (high-impurity)
tail end are sawed off. The exact crystallographic orientation of the ingot can be checked
by X-ray crystallography (usually not necessary). The ingot is milled into a square
cross section and sawed into individual wafers by a liquid-cooled diamond-studded
blade or steel wire. Internal diameter saws have been the mainstay of the industry. The
inner edge of a thin metal annulus serves as the cutting edge. At high rpm, the blade
becomes rigid. It slowly cuts through the ingot, one wafer at a time. A more recent
method is the continuous-wire saw in which a thin steel wire is wrapped between two
rollers so that the wire segments are approximately parallel to each other. The wire
segments grind a silicon carbide slurry through the ingot. In spite of much greater
initial cost, wire sawing is replacing internal diameter sawing because it allows several
hundred wafers to be cut simultaneously and significantly reduces material loss inherent in cutting, i.e., *kerf loss*. A variation of wire sawing that further minimizes kerf loss
is the use of a reciprocating parallel wire gang (Fixed Abrasive Slicing Technique)[25].
FAST is not yet widely commercialized. It uses wires with a shaped cross section optimized for cutting. Regardless of the method, after cutting, the individual wafers undergo chemical etching to remove surface damage and mechanical-chemical polishing
to render a highly flat specular surface suitable for processing.

B. Cast Polycrystalline - A Moderate Efficiency Material

Large-grain polycrystalline (semicrystalline) silicon ingots are produced by the directional solidification of molten electronic-grade (or near electronic grade) silicon in a
silica crucible with a rectangular cross section (fig. 2.4-2). The term *casting* is commonly used to refer to any process where the melt freezes in a mold. In some cast
processes, the liquid silicon is poured from the crucible into a silica mold. In other
versions, the melt is allowed to freeze in the crucible. Historicaly, the main advantage
of pouring was that it allowed a large mass of material to be processed, whereas the
one-container method had lots of dead space between the chuncks in the silicon charge.
This distinction is becoming vague because the largest commercial cast ingots as of
1996 – 240 kg and 66 cm square – are made without pouring[26]. Because of a phase

Fig. 2.4-2   Cast large-grain polysilicon (semicrystalline) ingot, quartered brick, wafer, and cells. (*Photo courtesy Solarex*).

change that occurs in solid silicon as it cools down, a mechanical mismatch develops between the silicon ingot and the mold or crucible. Consequently, the silica structure that holds the freezing melt fractures upon cooling. Regardless of the process, the temperature of the freezing melt is controlled by resistance heaters so that solidification occurs from the bottom up. An exception is the heat exchange method (HEM)[27]. In the heat exchange method, solidification from the bottom up is controlled by a gaseous heat exchange fixture at the bottom of the melt crucible. A major expense of the cast process is the sacrifice of the mold (or crucible). HEM is a non-pouring process that has been used with a reusable non-oxide ceramic crucible so that replacement crucible costs are minimized.

With any cast method, the resulting polycrystalline ingot has very large grains – typically 1 mm to 30 mm in diameter. The majority of the grains are about 20 mm across. Unintentional impurities with distribution coefficients well below 1 segregate into the melt during freezing and are finally deposited at the top of the ingot. The top is sawed off before the ingot is sliced into wafers. Boron is added to the melt to control resistivity. There is no need to mix the impurity into the melt. Convection forces stir the melt. This effect, along with the almost unity impurity distribution coefficient of boron, assure boron uniformity in the finished ingot. When the cast has completely solidified

Fig. 2.4-3   Semicrystalline silicon wafer, approximately 10-cm wide, sliced from an ingot. The large grains are vertically oriented along the direction of solidification. (*Photo courtesy Solarex*).

and cooled, the ingot is delaminated from the mold or crucible. Ingots commonly have cross sections over 30 cm on a side and masses over 100 kg.

A critical requirement of semicrystalline material is that the cast or melt is cooled slowly with minimal temperature gradients so that large *columnar* (vertically oriented) grains are produced along the direction of solidification. The ingot is sliced into bricks and then individual wafers. The resultant wafers are characterized by large grains that extend more or less orthogonally through the wafer (fig. 2.4-3). This characteristic of vertically oriented grains significantly lowers the degree of minority carrier recombination that would otherwise appear at the grain boundaries. Interestingly, the purity requirements for semicrystalline material are at least as severe as for single-crystal material. Because of the already existing defects, semicrystalline material has less tolerance for unintentional impurities. Semicrystalline cells are between 85% and 100% as efficient as similarly processed single-crystal cells, depending on the material and cell fabrication technology.

Aside from the issue of impurities, cast semicrystalline silicon has *two outstanding qualities* that makes it a good choice for solar cells. The casting process is less energy intensive than Czochralski or float-zone techniques. This alone makes it a low-cost alternative to single-crystal. The second advantage has to do with the geometry of the wafers. Square wafers are preferred over the circular wafers typical of Czochralski or float-zone growth because of improved module packing density. The cast ingot has a square cross section to begin with and rectangular bricks from the ingot can be further sawed into square wafers with a minimal loss of material. For example, an ingot 32 cm x 32 cm in cross section, can be sawed into four 16 cm x 16 cm bricks, which, in turn, can each be sawed into many thin (about 300 microns) 16 cm x 16 cm wafers. Square wafers can be obtained from a Czochralski or float-zone ingot by milling the ingot, but this results in huge material loss. Maximum experimental 1-sun AM1.5-spectrum efficiencies for cast semicrystalline silicon cells and modules are about 18% and 15%, respectively. However, most production line cast semicrystalline cells and modules have efficiencies closer to 13% and 12%, respectively. An efficiency over 15% has been demonstrated for a prototype module made from inexpensively-processed HEM-material cells. *In 1995, cast polycrystalline accounted for about 25% of worldwide shipments of photovoltaic modules.*

C. Polycrystalline Ribbon - An Alternative to Ingot Technologies

A crystalline alternative to the two ingot technologies is ribbon growth. Ribbon technologies produce long flat semicontinuous strips (ribbons) of large grain polycrystalline material several hundred microns thick and 5 to 10 centimeters across. Growth rates are not more than three centimeters per minute. These methods rely on the high surface tension of molten silicon that enables a long strip to be shaped as it is drawn out of a melt. Ribbon processes are semicontinuous because the liquid can be drawn out of the melt as long as a supply exists in the crucible. In some cases, the crucible is resupplied with feedstock while the ribbon is being drawn. The term "ribbon" can be somewhat misleading, as the thickness of the product prevents it from being wound on a spool like a cloth ribbon. Silicon does not become flexible until the thickness is decreased to about 6 microns. There are numerous variations to the ribbon growth approach, but they can be classified according to the mechanism for shaping the silicon as it is drawn from the melt. The ribbon can be edge shaped by vertically drawing filaments (strings) of refractory material through the silicon melt, shaped by a substrate that is pulled vertically through the melt, shaped by selective cooling as the melt is horizontally extruded from the melt, shaped by a graphite die as the melt is vertically pulled through the die by capillary force, or shaped by other methods. As of 1996, the most commercially viable of these techniques is the vertical *die shaping* approach. Growth is initiated by a seed. As soon as the melt exits the top of the die, it solidifies into polycrystalline material, and is mechanically grasped and pulled upward. The pulling speed and die geometry determine the thickness of the ribbon, typically 250-350 microns. When growth is terminated, the ribbon is divided into individual wafers by sawing or laser scribing with virtually no material loss.

About 2.5% of photovoltaic modules in 1995 were made from solar cells produced by ribbon growth technologies. While this is a small percentage of overall cell production, there are *three cost advantages* to this growth approach compared to the ingot technologies. First, and foremost, is the fact that very little silicon is lost through wafer sawing. Kerf loss is one of the major cost drivers of the ingot technologies because the lost silicon is expensive to grow in the first place. The loss depends on the characteristics and dynamics of the sawing procedure, not on the method for forming the ingot. Kerf loss is about 340 microns per wafer if an internal diameter blade saw is used[28]. This decreases to about 230 to 250 microns per wafer with a continuous wire saw, and as little as 175 microns per wafer with a shaped FAST wire[29]. However, this is still more than half the thickness of a typical wafer. The silicon dust is not suitable for reuse because of the contamination introduced by the wire slurry or blade coolant during the sawing. (Additionally, kerf has a tremendous surface area. This results in the formation of a large amount of SiO_x that makes the material unsuitable for future semiconductor use. Thus, kerf is discarded.) Since a ribbon needs only a perpendicular cut every ten to fifteen centimeters to form the individual wafers, little silicon is lost. Most of the silicon in the crucible goes into the final wafer product. Also, a ribbon is amenable to cutting by an infrared laser, and this speeds up the wafer separation process. Secondly, there is no surface damage due to sawing-ribbon wafers do not have to be mechanically or chemically polished (e.g., by etching off the top surface) and they do not have to be cleaned. Of the five major cost drivers in the production of crystalline silicon modules (material growth, wafer separation, cell fabrication, module materials, and module assembly and test), the ribbon approach almost completely eliminates one of these. Thirdly, ribbon technology has the (not yet realized) *potential* for high throughput per growing station: several thousand wafers per hour, versus about 350 wafers per hour for Czockralski material, 900 per hour for float-zone material, and 1000 per hour for cast material[30].

Ribbon technologies have had difficulties in reaching the market place because of high defect density and poor wafer planarity. Molten silicon is highly reactive and dissolves virtually all other materials. Graphite is chosen for the die because of its relatively slow dissolution rate. But for any die shaping system, dissolution of the graphite die by the melt produces a high defect density (by way of carbon impurities). The large graphite-contact-area to volume ratio of the shaping die causes a large amount of carbon to dissolve in the melt as it passes through the die. Additionally, as they solidify from the melt, ribbons have high cooling rates because of the high ribbon-area to volume ratio. This introduces mechanical stress and subsequent dislocations. The grain boundaries tend not to be orthogonal to the ribbon surface because of 2-dimensional cooling. The direction of heat leaving the ribbon has a component along the length of the ribbon as well as a component orthogonal to the surface. This characteristic alone tends to increase minority carrier recombination at grain boundaries. Die-induced carbon impurity and rapid cooling-induced defects tend to degrade the performance of ribbon cells compared to ingot-based cells. However, cell manufacturers have been able to overcome this shortcoming of ribbon material by making extensive use of impurity gettering and hydrogen passivation[31]. Passivation of surface and bulk defects by hydrogen diffusion induced by deposition of silicon nitride has been very effective in boosting performance[32]. For the most successful of the ribbons, edge-defined film-fed growth (EFG) material, oxygen incorporation through the addition of CO in the growth atmosphere has boosted cell efficiency 0.6% absolute so that cell efficiencies now routinely approach 14%[33]. This mechanism is not fully understood, but improvement of several cell parameters suggests a reduction of recombination rates throughout the cell. Diffusion lengths approach the thickness of the cell, about 300 μm.

High cooling rate also causes a ribbon to have poor planarity or flatness. This is significant because it increases the frequency of wafer cracking during the soldering and plastic encapsulation processes when the cells are arranged to form a module. Even if a crack is not severe enough to cause a piece of the wafer to fall off, it can induce a very low shunt resistance between the front and back side of the wafer which will ruin the cell's performance. For solar cells in general, modelling has shown a small region of poor material can drastically reduce the performance of the cell[34]. To prevent cracking, manufacturers of ribbon cells have had to develop refinements to the metal screen printing and polymer lamination procedures that accommodate the relatively poor planarity of the material.

REFERENCES

[1]   W.R. Runyan, *Semiconductor Measurements and Instrumentation*, pp. 21-42, McGraw-Hill Book Company, New York, 1975.

[2]   B.R. Bathey, et al., "On the use of oxygen as an upgrading element in high efficiency EFG solar cells," *Proc. of the 11th E.C. Photovoltaic Solar Energy Conf.*, pp. 462-464, Oct. 1992.

[3]   W. Shockley, *Electrons and Holes in Semiconductors*, Chap. 7, Van Nostrand Reinhold Co., New York, 1950.

[4]   W. Shockley and W.T. Read, "Statistics of the recombination of holes and electrons," *Phys. Rev.*, vol. 87, pp. 835-842, Sep. 1952.

[5]   C.T. Sah, R.N. Noyce, and W. Shockley, "Carrier generation and recombination in p-n junctions and p-n junction characteristics," *Proc. IRE*, vol. 45, pp. 1228-1243, Sep. 1957.

[6]   W.M. Bullis, "Properties of gold in silicon," *Solid State Elec.*, vol. 9, pp. 143-168, 1966.

[7]   A.K. Jonscher, *Principles of Semiconductor Device Operation*, pp. 23-27, G. Bell & Sons, London, 1960.

[8]   A.S. Grove, *Physics and Technology of Semiconductor Devices*, pp. 106-113, John Wiley and Sons, New York, 1967.

[9]   N.W. Ashcroft and N.D. Mermin, *Solid State Physics*, pp. 562-565, Holt, Rinehart and Winston, Inc., Philadelphia, 1976.

[10]   F.A. Trumbore, "Solid solubilities of impurity elements in germanium and silicon," *Bell Sys. Tech J.*, vol. 39, pp. 205-234, Jan. 1960.

[11]   A. van der Ziel, *Solid State Physical Electronics*, pp. 49-57, Prentice-Hall Inc., Englewood Cliffs, NJ, 1976.

[12]   F.A. Lindholm and C.T. Sah, "Fundamental electronic mechanisms limiting the performance of so-lar cells," *IEEE Trans. on Electron Devices*, vol. ED-24, pp. 299-304, Apr. 1977.

[13]   R.K. Brenneman, et al., "Issues in solar-grade silicon feedstock development," *11th E.C. Photovoltaic Solar Energy Conf.*, pp. 412-415, Oct. 1992.

[14]   W. Keller, *Floating-Zone Silicon*, pp. 3-9, Marcel Dekker, New York, 1981.

[15]   H. Schweickert, H. Gutsche, and R. Emeis, U.S. Patent # 3,030,189, April 17, 1962.

[16.]   Private communication, R. Brenneman, May 22, 1996.

[17]   Y. Sakaguchi, et al., "Metallurgical purification of metallic grade silicon up to solar grade," *Proc. of the 12th E.C. Photovoltaic Solar Energy Conf.*, pp. 971-974, Apr. 1994.

[18]   F. Ferrazza, et al., "Basic requirements for solar grade silicon feedstock," *Proc. of the 12th European Photovoltaic Solar Energy Conf.*, pp. 1007-1008, Apr. 1994.

[19]   R.L. Lane and J. Boothroyde in *Silicon Processing for Photovoltaics I*, ed. by C.P. Khattak and K.V. Ravi, Chap. 2, North Holland Physics Publishing, New York, 1985.

[20]   Product literature for Kayex-Hamco KX-150 Crystal Growing Furnace, May 1996.

[21]   J.R. McCormick in *Silicon Processing for Photovoltaics I*, ed. by C.P. Khattak and K.V. Ravi, Chap. 1, Table 5, North Holland Physics Publishing, New York, 1985.

[22]   W. Keller and A. Muhlbauer, *Floating-Zone Silicon*, Chap. 1, Marcel Dekker, Inc., New York, 1981.

[23]   W.C. Dash, *Growth and Perfection of Crystals*, ed. by R.H. Doremus, B.W. Roberts, and D. Turnbull, John Wiley and Sons, New York, p. 361, 1958.

[24]   T.H. Wang, T.F. Ciszek, and T. Schuyler, "Micro-defect effects on minority carrier lifetime in high purity dislocation-free silicon single crystals," *Solar Cells*, vol. 24, pp. 135-145, 1988.

[25]   F. Schmid, M.B. Smith, and C.P. Khattak, "Development of shaped wire for FAST slicing," *Proc. of the 12th European Photovoltaic Solar Energy Conf.*, pp. 1009-1010, Apr. 1994.

[26]   Product literature by Crystalox Limited, Oxfordshire, U.K., May 1996.

[27]   C.P. Khattak, et al, "Characteristics of HEM silicon produced in a reusable crucible," *Conf. Rec. of the 23rd IEEE Photovoltaics Specialists Conf.*, pp. 73-77, May 1993.

[28]   T.L. Jester, et al., "Photovoltaic CZ silicon manufacturing technology improvements," *Conf. Rec. of the First World Conf. on PV Energy Conversion*, pp. 824-827, Hawaii, Dec. 1994.

29   F. Schmid, M.B. Smith, and C.P. Khattak, "Method for kerf reduction in fast slicing," *Proc. of the 11th E.C. Photovoltaic Solar Energy Conf.*, pp. 477-479, Oct. 1992.

30   Notes from panel discussion on silicon process throughput, *Fifth NREL Conf. on the Role of Impurities and Defects in Silicon Device Processing*, Copper Mountain, CO, Aug. 1995.

31   D.S. Ruby, et al., "Optimization of PECVD deposition processes for commercial multicrystalline silicon solar cells," *Abstracts of the 25th IEEE Photovoltaic Specialists Conf.*, no. 150, May, 1996.

32   P. Doshi, et al., "High efficiency large and small area Si solar cells using low-cost rapid thermal processing, screen printing, and plasma-enhanced chemical vapor deposition," preprint of paper presented at 25th IEEE Photovoltaic Specialists Conf., May 1996.

33   B.R. Bathey, et al., "On the use of oxygen as an upgrading element in high-efficiency EFG solar cells," *Proc. of the 11th E.C. Photovoltaic Solar Energy Conf.*, pp. 462-464, Oct. 1992.

34   F.A. Lindholm, et al., "Degradation of solar cell performance by areal inhomogeneity," *Solid State Electronics*, vol. 23, pp. 967-971, Sept. 1980.

3

SOLAR CELL MECHANISM AND PERFORMANCE

A solar cell is a large-area pn-junction (pn-diode) structure designed for efficient conversion of sunlight into electric current. Solar cells employ the *photovoltaic effect* whereby excess photogenerated minority carriers are separated by a junction with a built-in field. Once they are separated into opposite sides of the junction, they become majority carriers. It is the excess concentrations of these *majority* carriers that creates a voltage across the external circuit. Current passes through some load in the external circuit to do useful work. Electrons are forced out through one terminal of the diode, and holes are forced out through the other. At any given point in the circuit, the sum of the hole and electron currents is the *total* current of the solar cell. The positive direction of current is that of holes into the device. Consequently, the current from the device is negative and the I-V characteristic of the cell appears in the fourth quadrant of the I-V plane, indicating work is being done on the load.

3.1 STRUCTURE OF THE GENERIC SOLAR CELL

The generic pn-junction solar cell is a monolithic structure with two active layers: a thin heavily doped top layer or *emitter*, and a thick, moderately doped, bottom layer or *base*. For conventional silicon cells made from bulk material, the emitter is about 0.3 to 0.8 μm thick and the base is 100 to 300 μm thick (fig. 3.1-1). The lower thickness limit is reserved for space cells. Base doping resistivities are usually in the range of 1-3 Ω cm.

The emitter and base have opposite dopant species. In a silicon cell, the emitter is doped n-type with an average concentration of approximately $10^{18}/cm^3$, and the base is doped p-type with a *uniform* concentration of approximately $10^{16}/cm^3$. The emitter is formed by making a high-temperature diffusion from a gaseous or liquid source into the base material. Its concentration varies with depth. Surface concentration can be greater than the solid solubility of the dopant, but steadily decreases with depth until the concentration of n-type dopant equals the concentration of p-type dopant. At that

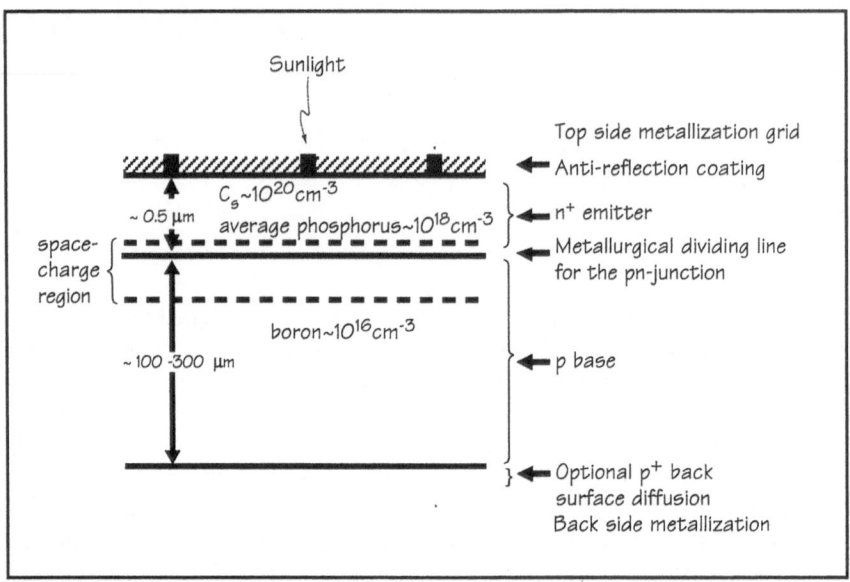

Fig. 3.1-1   Cross section of the common silicon pn-junction solar cell.

depth, the material changes from n-type to p-type, and the interface is referred to as the *metallurgical* junction. A pn-junction with its built-in electric field is formed in the vicinity of the metallurgical junction. The width of the space-charge region (SCR) is about 1 μm, with most of the SCR on the lesser-doped p-side of the metallurgical junction.

The bottom surface of the cell is covered with metal and alloyed (heat treatment above the Al-Si eutectic temperature of 580°C) to form an Al-Si solution on the back surface. This creates a non-rectifying (ohmic) contact. Top side electrical contacts take the form of thin fingers, rather than complete metallization, so that only a small percentage of the cell's top surface (3-10%) will be shaded when exposed to sunlight. However, cells intended for concentrated light require both lower top metal resistance and lower top surface resistance to accommodate the high current in the circuit. Otherwise, the series resistance becomes prohibitive. This mandates either thicker or broader metallization fingers, and fingers which are closer together. Top metal shading for concentrators is 15% to 20%. The top surface is moderately conductive because it is heavily doped. This yields a sheet resistance of between 30 and 100 $\Omega/\square$ and allows the formation of an ohmic contact. Before metallization, the emitter surface is passivated with an oxide or nitride, SiO_2 or SiN_x, to minimize surface recombination velocity. These thin films, as well as TiO_2, also serve as anti-reflection coatings. If the anti-reflection coating is separate from the surface passivant, it can be applied either before or after metallization, depending on the process.

3.2 PHOTOGENERATED CURRENT

In a closed solar cell circuit, the current path involves some external device, e.g., an electric motor or storage battery. The solar cell must meet *three fundamental requirements* to supply power to the external device:

(i)   the cell must have a volume in which *excess* hole-electron pairs are generated by the absorption of sunlight;

(ii) there must be a mechanism for separating excess minority carriers and forcing them over to the other side of the junction, where they become majority carriers and are forced out through the contacts by Coulomb force; and,

(iii) the excess minority carriers must survive (avoid recombination) long enough to be separated by that mechanism.

The first requirement is met by providing a sufficiently long path length, either by designing a base thickness that exceeds the absorption length over a large portion of the usable spectrum or by employing a light trapping scheme that produces multiple passes through the bulk of the cell. For a semiconductor sample of any thickness, a photon of light is absorbed only if the energy of the photon is approximately equal to or greater than the bandgap value, E_g. However, there is some absorption of photons with energies just below the bandgap. Wavelength λ and photon energy E are related by $E = hc/\lambda$, where h is Planck's constant (6.63×10^{-34} Joule-sec or 4.14×10^{-15} eV-sec) and c is the speed of light in vacuum (3.00×10^8 meters/sec). Consequently, the wavelength must be approximately equal to or shorter than hc/E_g for the photon to be absorbed. For silicon, this imposes a limit on maximum wavelength of about 1.2 μm, which is about 0.1 eV below the bandgap; for GaAs, about 0.9 μm. If this requirement is met, the light is absorbed, with the intensity exponentially decaying with distance as it travels into the semiconductor. The exponential decay constant (called the skin depth or absorption length) is the distance in which the intensity falls to 1/e (about 37%) of its initial value. If the wavelength requirement is not met, the light is not absorbed at all, and passes through the semiconductor. Light can be absorbed by the metallization at the back surface, but this does not produce electron-hole pairs.

Absorption length is dependent on wavelength, varying in silicon from 0.7 μm at
$\lambda = 0.350$ μm (near-ultraviolet) to 1 cm at $\lambda = 1.100$ μm (near-infrared). Thus, the *blue
end of the spectrum is absorbed near the top of the solar cell*. On the other hand, much
of the light at the red end of the spectrum reaches the backside of the cell without being
absorbed in the semiconductor. Unless the red frequencies are reflected back into the
semiconductor by a backside coating or backside textured surface, they will pass through
the cell. Bulk silicon cells are designed to between 100 and 300 μm thickness to allow
an adequate pathlength for absorption. Cells can be made with very thick bases, but
there is a twofold trade-off between thickness and performance. When the base is more
than 400 μm thick, there is a requirement for greatly improved minority carrier lifetime
to allow collection by the junction. This implies use of expensive float-zone material.
Secondly, a thick base means fewer cells are obtained from each ingot. (By comparison, GaAs and other indirect bandgap crystals have very short absorption lengths and
cells made from these materials need only several microns thickness.) An alternative
approach to thick bulk material for achieving sufficiently long path length is light trapping. The front side of a cell made in (100) oriented material is pyrimidally textured
with KOH, NaOH, or other anisotropic etch to expose (111) planes. This is used primarily as an anti-reflection mechanism. A side effect is the oblique entry angle of light
into the semiconductor which increases the effective path length and also increases the
number of pairs produced near the junction. Additionally, a specular backside due to
polishing or aluminum film will reflect light toward the front surface and effectively
double the path length.

The second fundamental requirement is met by a pn-junction and the accompanying
built-in electric field (fig. 3.2-1). It is easy to see how a pn-junction can separate out
excess minority carriers that approach the metallurgical junction. If an excess electron
on the p-side, or an excess hole on the n-side, comes near the built-in electric field of
the space-charge region (fig. 2.3-4(a)), the field will immediately sweep the carrier
over to the other side. The carrier can be thought of as falling "downhill", energy-wise.
The junction separates out the excess electrons in the p-type base and the excess holes
in the n-type emitter. However, because the junction is close to the surface, most of the
light is absorbed in the base. Thus, most of the separated charge consists of excess
electrons from the base.

Use of a pn-junction, or any type of junction for that matter, is not necessary for separating charge carriers and producing a voltage at the terminals. In principle, these func-
tions can be achieved just by the illumination of one side of an intrinsic silicon sample
through the *Dember* effect. The unequal diffusivities and lifetimes of carriers produces
a voltage difference in the direction of the illuminated side. If intrinsic silicon is illumi-

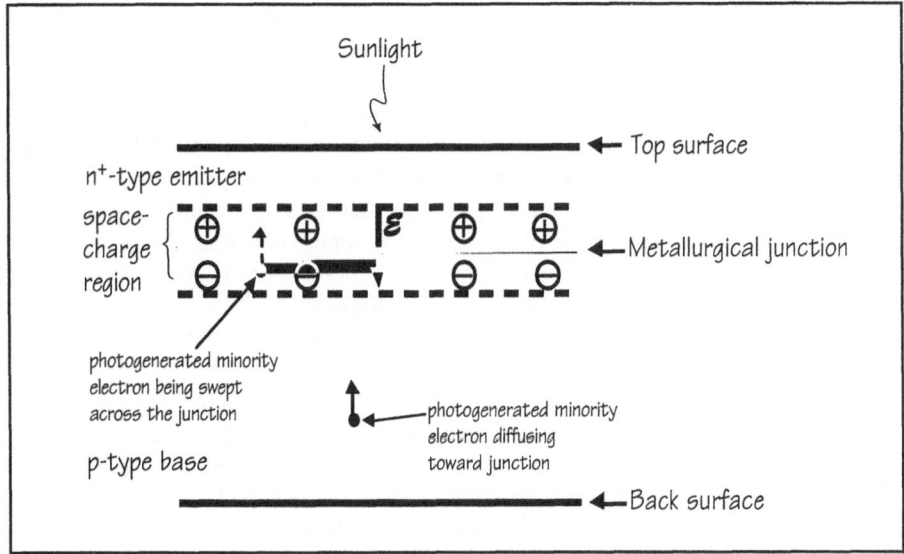

Fig. 3.2-1   The pn-junction separates the photogenerated (excess) minority carriers.

nated, the larger diffusivity of electrons coupled with decreasing absorption into the material produces a separation of holes and electrons. Hole concentration increases near the illuminated side, while electron concentration increases farther down in the sample. The electric field produced by this separation is weak. Open-circuit voltages are on the order of a few millivolts, rather than the approximately 600 mV seen in pn-junction commercial cells.

The third requirement is the most problematic. It is nominally met by moderately doping the base (1-3 Ω cm) and using material with low crystallographic defect density. With these qualities, the diffusion length for excess electrons in the base will be large (> 100 μm). The pn-junction can be thought of as a sink for the excess photogenerated minority carriers. Under steady-state illumination, excess photogenerated electrons are constantly formed in the bulk of the base. They disappear from the base as they approach the junction, and produce a concentration gradient for electrons. The gradient maintains diffusion of excess minority carriers toward the junction and is maximized by a large built-in voltage that assures the rapid disappearance of those carriers at the edge of the space-charge region. Thus, increased base doping serves the purpose of providing a strong built-in voltage at the junction. This tends to increase the open-circuit voltage of the cell since V_{oc} (discussed below) is *limited* by the built-in voltage.

On the other hand, the survivability of excess minority carriers in the base *decreases* with increased boron doping. This makes an argument for lower base doping. Overall, the issue of collectibility of excess minority carriers and optimal base resistivity is a trade-off between increased built-in voltage and increased survivability in the bulk. Traditionally, transport of excess carriers to the junction has been thought to be dependent on recombination mechanisms. Recent work suggests the possibility that transport mechanisms may be independent processes that control recombination[1, 2]. This possibility is consistent with the observation that some very high efficiency silicon cells display non-ideal diode current-voltage (I-V) curves. For such cells, there is very little base recombination in the first place. High efficiency is attributable to non-linear transport effects on the I-V curve rather than to lowered recombination rates in the bulk[3].

Related to the issue of collectibility is the problem with photogenerated minority carriers that wander toward the back contact. In some solar cells, a p^+/p junction is intentionally formed on the backside to create an electric field that turns around any excess electrons that diffuse toward the back surface. This is a *back-surface field*. At the cost of additional processing, it can significantly improve the collectibility of excess minority carriers and the subsequent efficiency of the cell.

Thus, a solar cell uses its thick base to absorb sunlight and create excess electrons, which, if they can avoid recombination long enough, will diffuse over to the pn-junction. The junction separates out the excess electrons from the base and sends them over to the emitter. Once in the emitter, they become *majority* carriers, and feel little concentration gradient. However, the *excess* electrons in the emitter tend to repel each other and the electrostatic force (Coulomb force) pushes them out the top contact of the cell. Coulomb force tends to smooth out the excess majority carrier charge, a process characterized by the dielectric relaxation time in the emitter. It is the photogenerated excess charge, and *not the built-in voltage*, that produces a voltage across the external circuit.

The *excess* holes in the emitter behave analogously to the excess electrons in the base. There are several quantitative differences, though. Because the junction is so close to the top surface, most of the light reaches the base, and few electron-hole pairs are created in the emitter. Those pairs that are created in the emitter are produced by the blue end of the spectrum, which has a very short absorption length. The excess holes in the emitter tend to diffuse toward the junction where they get swept into the p-type base to become majority carriers. Once they become majority carriers, Coulomb force pushes them out through the base contact on the backside of the cell. Few holes in the emitter survive long enough to make it all the way to the junction. This is because the diffusion length in the emitter is very short as a consequence of the high doping concentration. Most of the photogenerated current in a (n^+/p silicon) solar cell circuit consists of excess electrons that are created in the base. The diffusion length in the emitter could be improved by lowering the doping density there, but eq. (2.3-7) indicates this would decrease the strength of the built-in voltage and the subsequent upper limit for the open-circuit voltage of the cell.

From the above description, it might appear that the current in a closed solar cell circuit during illumination is merely the photogenerated current. Unfortunately, this is not the case.

3.3 DARK RECOMBINATION CURRENT OPPOSES PHOTOGENERATED CURRENT

During illumination, the pn-junction separates the excess photogenerated *minority* carriers from the quasi-neutral regions and transports them to the opposite sides of the junction, where they become *majority* carriers. However, the presence of these excess majority carriers near the edges of the space-charge region has the effect of biasing the pn-junction. This produces the same effect as if the solar cell was forward biased by a battery.

Figure 2.3-4(b), shows that the application of a forward bias lowers the barrier height of the junction, and allows an increase in the rate at which *majority* carriers diffuse across the junction and become minority carriers. Thus, the separation of photogenerated charge has the effect of *forward* biasing the junction and causing the formation of a current that *opposes* the photogenerated current (fig. 3.3-1). The built-in voltage is an upper limit for the forward bias across the junction. In an n^+/p cell, while the excess photogenerated electrons from the base flow out of the emitter terminal, the forward-biased junction injects electrons from the emitter back into the base. (Likewise, the forward bias injects holes from the base into the emitter.) The injection current is equal to the current in a forward-biased diode in the dark (fig. 2.3-5(a)); thus, it is called the *dark diode current*.

Bias across the junction changes the minority carrier concentrations at the edges of the space-charge region. From eq. (2.2-9), with the approximation that there is no recombination or generation in the SCR, the np product throughout the SCR is

$$n\ p = n_i^2\ e^{q(V_P - V_N)/kT} \tag{3.3-1}$$

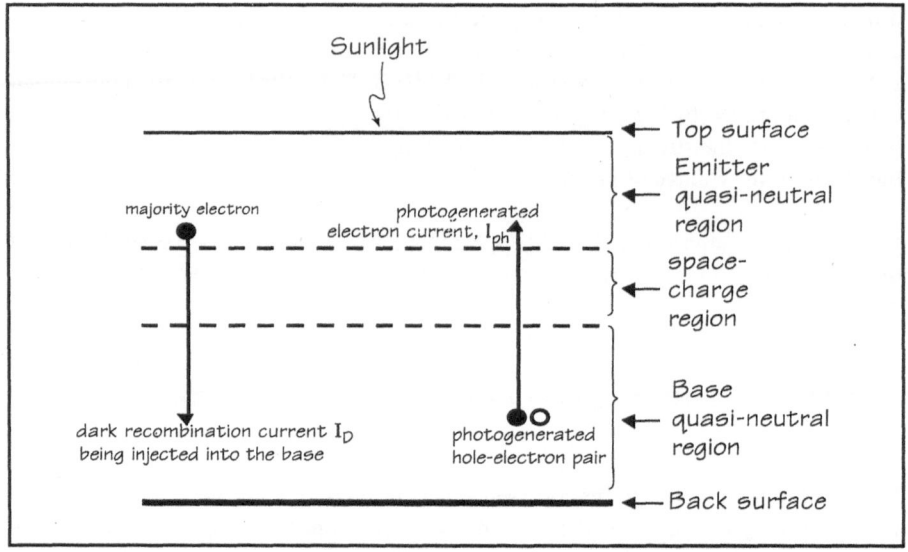

Fig. 3.3-1   There are two currents in a pn-junction solar cell: photogenerated current and dark recombination current.

With the assumption that all of the bias is dropped across the SCR, the quasi Fermi levels can be taken as flat across the SCR and eq. (3.3-1) becomes

$$n \ p = n_i^2 \, e^{qV/kT}$$

where V is the bias. At the p-edge of the SCR (i.e., x = 0), the minority electron concentration is then

$$n(0) = \frac{n_i^2}{p(0)} \, e^{qV/kT}$$

With the assumption of low injection, p(0) is approximately equal to the equilibrium hole concentration, i.e., $p(0) \approx p_{po}(0)$, where the subscript denotes p-type material and thermal equilibrium. Then

$$n(0) = \frac{n_i^2}{p_{po}(0)} \, e^{qV/kT}$$

$$= n_{po}(0)e^{qV/kT} \qquad (3.3\text{-}2)$$

The *excess* electron concentration at the p-edge of the SCR is then

$$\Delta n(0) = n_{po}(0)e^{qV/kT} - n_{po}(0)$$

$$\Delta n(0) = n_{po}(0)(e^{qV/kT} - 1) \qquad (3.3\text{-}3)$$

There is a similar expression for excess holes at the n-edge of the SCR. For forward bias, the excess minority carrier concentrations at the SCR edges are much greater than their equilibrium values (fig. 3.3-2). For a reverse-bias diode, the minority concentrations at the SCR edges approach zero.

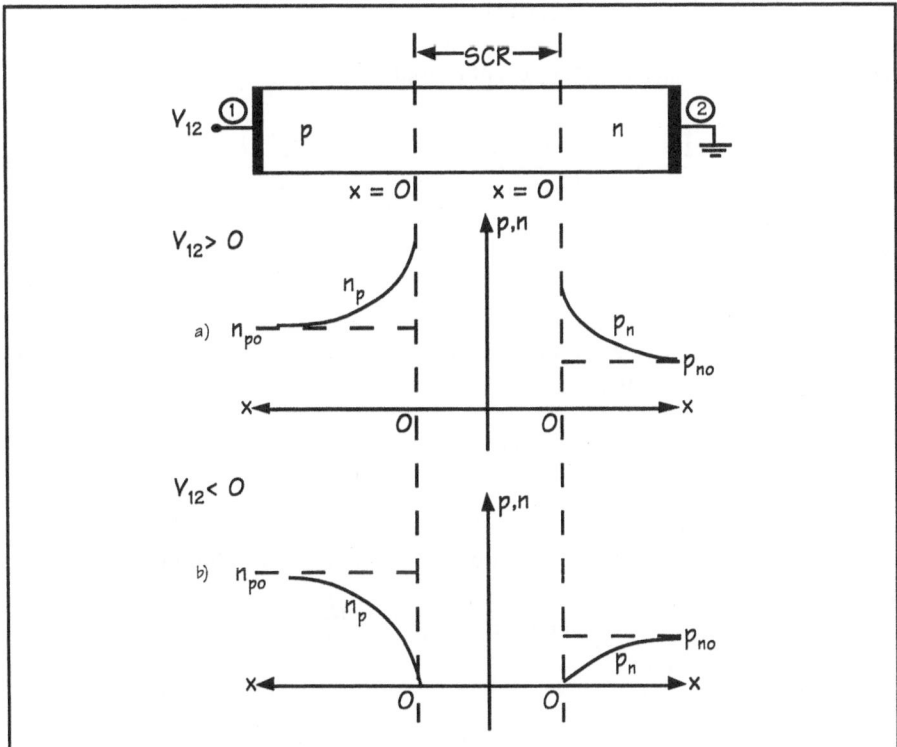

Fig. 3.3-2   Minority carrier concentrations at the edge of the SCR. (a) Forward bias. (b) Reverse bias.

Diffusion length has been defined as the mean length for minority carrier recombination. In terms of the excess electron concentration at the p-edge of the SCR, there is now an alternative physical interpretation for L as the exponential decay length for minority carriers. Assume, for a moment, that this is the case for electrons injected into the base. Then at a distance x into the base QNR,

$$\Delta n(x) = \Delta n(0)e^{-x/L_n} \qquad\qquad (3.3\text{-}4)$$

The probability that an electron injected at x = 0 survives to a distance x is

$$\frac{\Delta n(x)}{\Delta n(0)} = e^{-x/L_n}$$

The probability that an electron at x will recombine at dx is

$$\frac{\Delta n(x) - \Delta n(x + dx)}{\Delta n(x)} = \frac{-d\Delta n(x)}{dx} \frac{dx}{\Delta n(x)}$$

$$= dx / L_n$$

By taking the product of the above two probabilities, the probability that an electron injected at x=0 will recombine in a given dx is

$$e^{-x/L_n} \frac{dx}{L_n}$$

Then the median distance that an electron diffuses before recombining is

$$\langle x \rangle = \int_0^\infty x e^{-x/L_n} \frac{dx}{L_n}$$

$$= L_n$$

Thus, diffusion length L_n is *both* the median length to recombination and the exponential decay length.

For forward bias, the injected dark electrons start diffusing away from the p-edge of the SCR in response to their own large concentration gradient. At the edge of the space-charge region, the excess minority carrier concentration is equal to its thermal equilibrium value times the factor exp(qV/kT)-1. But far into the quasi-neutral base, the injected electron concentration will be zero. After travelling a few diffusion lengths, the injected electrons almost completely disappear as they recombine with the surrounding majority (hole) carriers, or they reach the back contact and disappear by flowing into the metal. It is for this reason that the dark current is often referred to as the *dark recombination current* of the solar cell. Notice, for those minority carriers reaching the vicinity of the back contact, if the rate of disappearance at the back contact can be decreased, i.e., a decreased back surface recombination velocity, then the concentration gradient is lessened. In this case, the dark (injected) carrier current will be lowered. Thus, a back surface p$^+$ layer (back surface field) is expected to increase the collectibility of photogenerated carriers as well as decrease the dark current.

The dark current in a solar cell is *strictly parasitic*. It detracts from the photogenerated current, and lowers the net current in the circuit and the efficiency of the cell. The magnitude of the dark current depends on the rate at which injected carriers can recombine in the quasi-neutral regions. For a steady state condition, carriers will only be injected as fast as they can recombine in the bulk of the base. Thus, to lower this parasitic current component, the minority carrier diffusion length, particularly in the base, must be made as large as possible. (Alternatively, the base can be made thinner.) This partly explains the relatively high efficiencies of cells made from single-crystal or large-grain polycrystalline silicon. For these materials, the diffusion length in the base is usually several hundred microns and the dark current is relatively small. Additionally, a large diffusion length will increase the photogenerated current by enabling photogenerated minority carriers in the base to diffuse to the vicinity of the junction and be separated. Much of the research in silicon solar cells is oriented toward improving the minority carrier diffusion length.

An analytical expression for the steady-state dark recombination current is readily obtained with six simplifying assumptions. These prove to be reasonable for many crystalline silicon cells:
   (1) the quasi-neutral base is uniformly doped;
   (2) the base is much longer than the diffusion length;
   (3) all of the bias voltage is dropped across the space-charge region;
   (4) low-injection;
   (5) there is no recombination or generation in the SCR; and
   (6) the current is one-dimensional.

With the initial assumption of no recombination or generation in the space-charge region, the dark current consists only of a diffusion component in the quasi-neutral regions. Consequently, the dark current in each QNR is given by the solution to a diffusion equation. The boundary conditions are set by the excess minority carrier concentrations at the extremities of the QNR.

Consider the p-type base QNR. At any given point, the rate of change of minority carriers equals the net inward flux at that point, plus the contribution due to generation, minus the contribution due to recombination. This is expressed by the *continuity equation* for minority electrons in the base:

$$\frac{\partial n}{\partial t} = \frac{-\nabla \cdot J_n}{-q} + (G - R)$$

Here $J_n = q\, D_n\, \nabla n$, $G = 0$, and n can be written as $n = \Delta n + n_{po}$. For low-injection, the recombination rate R is proportional to the excess minority carrier concentration: $R = \Delta n / \tau_n$. With these substitutions, the one-dimensional equation is

$$\frac{\partial n}{\partial t} = D_n \frac{d^2}{dx^2}(\Delta n + n_{po}) - \frac{\Delta n}{\tau_n}$$

In the base QNR, n_{po} is constant. For steady-state, all time derivatives are zero. This reduces the equation to

$$\frac{d^2}{dx^2}(\Delta n) - \frac{\Delta n}{D_n \tau_n} = 0$$

$D_n \tau_n = L_n^2$ yields

$$\frac{d^2}{dx^2}(\Delta n) - \frac{\Delta n}{L_n^2} = 0 \qquad (3.3\text{-}5)$$

From eq. (3.3-3) and the initial assumption of a long base, the boundary conditions are

$$\Delta n(0) = n_{po}(e^{qV/kT} - 1) \qquad \Delta n(\infty) = 0 \qquad (3.3\text{-}6)$$

The solution to eq. (3.3-5) is then

$$\Delta n(x) = n_{po}(e^{qV/kT} - 1)\, e^{-x/L_n} \qquad (3.3\text{-}7)$$

This agrees with eq. (3.3-4) which is based on the physical interpretation of L_n. The electron current density in the base is then

$$J_n(x) = qD_n \frac{d}{dx} \Delta n$$

$$= \frac{-qD_n n_{po}}{L_n}(e^{qV/kT} - 1)\, e^{-x/L_n}$$

For diode cross-sectional area A, the current $I_n = AJ_n$. From the Law of Mass Action, $n_{po} = n_i^2/p_{po} = n_i^2/N_{AA}$, where N_{AA} is the base doping concentration. With these substitutions, the dark electron current magnitude at the *p-edge* of the SCR (x=0) is

$$I_n(0) = \frac{qAD_n n_i^2}{L_n N_{AA}}(e^{qV/kT} - 1) \qquad (3.3\text{-}8)$$

By a similar argument, the hole diffusion current at the *n-edge* of the SCR (x=0) is

$$I_p(0) = \frac{qAD_p n_i^2}{L_p N_{DD}}(e^{qV/kT} - 1) \qquad (3.3\text{-}9)$$

Note that x=0 refers to the p-edge or n-edge of the SCR, depending on the context. Equation (3.3-9) is the bulk recombination component in the emitter. However, the initial assumptions tend to be less valid in the emitter because the dopant concentration varies rapidly.

Current continuity requires the *total current* in steady state to be the same for any point in the diode. For convenience, take that point to be the p-edge of the SCR. The assumed absence of generation or recombination in the SCR implies that I_n and I_p are constant throughout the SCR. Thus, the total dark current is the sum of the above two currents:

$$I_{total\ dark} = I_n(0) + I_p(0) \qquad (3.3\text{-}10)$$

For most crystalline silicon cells, the base recombination current dominates the dark current, and is a good approximation to the total dark current. The current in eq. (3.3-10) is an ideality. As noted in [3], high efficiency silicon cells show non-ideal I-V curves due to non-linear processes. As an example, an assumption implicit in eq. (3.3-10) is that the base minority carrier mobility (diffusivity) is uniform. If this parameter is dependent on the voltage and carrier densities, then the assumption is violated and a non-linear process is introduced.

One might ask, how does a minority carrier "know" whether it is a dark injected carrier or a photogenerated carrier, and thus, how does it know which way to diffuse in the QNRs? The answer is simply that the problem is being divided up into two separate problems, the solutions of which are then *superimposed*. The photogenerated carriers feel a concentration gradient caused by the combination of light-induced carrier generation in the QNRs and carrier disappearance at the junction, and thus, diffuse toward the junction. The dark injected carriers feel a concentration gradient caused by the combination of carrier injection at the junction and carrier disappearance at recombination sites in the QNRs, and thus, diffuse away from the junction. For crystalline silicon cells, the total *illuminated* current is the combination of these two components[4]. More precisely, a solar cell may be thought of as a system with two inputs and two outputs. The inputs are the optical generation rate in the base and the excess minority carrier concentration at the edge of the space-charge region in the base (if the recombination current in the quasi-neutral emitter is negligible). The corresponding outputs are the short-circuit photocurrent and the dark recombination current, respectively. It is shown in[5] that the shifting approximation is valid if this system is *linear*. In[6] it is shown that the superposition principle may remain practically valid despite some nonlinearity in the system.

3.4 EQUIVALENT CIRCUIT MODEL

The equivalent circuit diagram models the solar cell and its external load as a lumped parameter circuit. In fig. 3.4-1, the photogeneration mechanism is represented by the current generator on the left side of the circuit. The dark recombination current mechanism is represented by the diode. The magnitudes of their currents are I_{ph} and I_D, respectively. Note that the current generator and diode are oriented in opposite directions

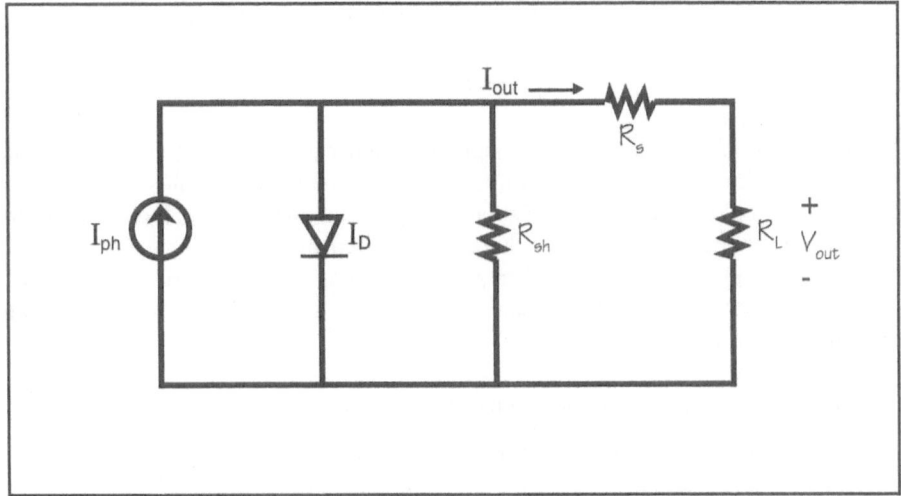

Fig. 3.4-1 Equivalent-circuit model of the solar cell. I_{ph} is photogenerated current, I_D is dark current, R_s is series resistance, and R_{sh} is shunt resistance, and R_L is the load.

since the diode current detracts from the photogenerated current. The equivalent circuit model has three resistors. The resistor R_{sh} represents any parallel high-conductivity paths (shunts) through the solar cell or on the edges of the cell. These would be caused, for example, by crystal damage in the junction, or a metallization spike through the junction. The resistor R_s represents the series resistance in the top surface of the semi-conductor and at the metal contact-to-semiconductor interface. The external load is R_L. The current through the load is I_{out}, and the voltage across the load is V_{out}.

R_{sh} and R_s are strictly parasitic. A low shunt resistance will degrade V_{out}, and a high series resistance will degrade I_{out}. In a good silicon cell, R_{sh} will be at least 500 Ω, and R_s will be less than 0.5 Ω.

Two special cases of interest in the equivalent circuit diagram are a short circuit across the load and an open circuit. If the load is shorted, the output current is the *short circuit current* I_{sc}. If the circuit is opened, the voltage across the diode and generator is the *open circuit voltage* V_{oc}. Note that if R_{sh} and R_s are negligible, i.e., R_{sh} very large and R_s very small, I_{ph} will equal I_{sc}.

Though R_{sh} and R_s are seldom negligible, it is useful to consider the *idealized case* where R_{sh} is very large and R_s is very small. The output current is then given by

$$I_{out} = I_{SC} - I_D \qquad (3.4\text{-}1)$$

(Here the currents are interpreted as magnitudes rather than algebraic values.) This is the superposition relationship or *shifting approximation* mentioned in section 3.3. Its use greatly simplifies the analysis of solar cell operation. In particular, it enables one to determine the current-voltage (I-V) characteristic simply by measuring the I-V characteristic of the cell in the dark, and then shifting that curve by the short circuit current. The determination of the dark I-V characteristic and the short circuit current are both straight-forward measurements.

The dark current of a *real* silicon cell has three distinct current components related to

(i)   recombination of injected carriers in the two quasi-neutral regions,

(ii) recombination in the junction space-charge region, and

(iii) the shunt current that leaks through the diode or around the edges of the cell

$$I_D = I_{QNR} + I_{SCR} + V/R_{sh}$$

For large shunt resistance, dark current is approximated by the QNR and SCR components

$$I_D = I_{QNRO} (e^{qV/1kT} - 1) + I_{SCRO} (e^{qV/2kT} - 1)$$

where the first term is given by eqs. (3.3-8) and (3.3-9), and the second term is derived in[7]. The pre-exponential for the SCR component is approximated by

$$I_{SCRO} = qAn_i W_{SCR} D_n / 2L_n^2$$

where W_{SCR} is the width of the space-charge region, and several simplifying assumptions have been made for the SCR.

(i)   There is a spatially uniform monoenergetic recombination-generation center at the middle of the bandgap;

(ii) hole/electron diffusivities are equal throughout the SCR; and

(iii) low-injection.

For any crystalline silicon cell, the QNR component is always evident when the I-V characteristic is plotted. When the QNR component is dominant for all values of V, the ln I vs. V plot is a straight line. Otherwise, for a non-negligible SCR component, there is a clear inflection point in the plot, and the exponent 2 will make the SCR component dominant in the low voltage region. By subtracting out the low voltage component from the total I-V plot, the QNR component can be graphically isolated[8]. If the QNR component is primarily from the base, it is approximated by eq. (3.3-8) in the case of large V. (For large V, the term "-1" is negligible.) Thus, the base QNR saturation current is obtained by extrapolating the QNR component to zero voltage

$$I_{QNRO} \approx I_{QNRO\ base} \approx \frac{qAD_n n_i^2}{L_n N_{AA}} \qquad (3.4\text{-}2)$$

Then eq. (3.4-2) allows the calculation of the base diffusion length L_n from the graphically extrapolated value of $I_{QNRO\ base}$.

Empirically, for large R_{sh}, the QNR and SCR components are described by a single term

$$I_D = I_O \left(e^{qV/n\ kT} - 1 \right) \qquad (3.4\text{-}3)$$

where I_o is the saturation current of the diode and n is a curve-fitting factor that (along with I_o) accounts for the combined effects of recombination in the quasi-neutral base and emitter regions and the space-charge region. If the dark current is dominated by recombination in the quasi-neutral regions, then n is exactly equal to 1. If there is also significant recombination in the space-charge region, then n is between 1 and 2. The value of n is also affected by the degree of light concentration. Under high concentration, minority and majority carrier densities become almost equal (high injection) in a region close to the junction. Values of n greater than 3 have been measured for shallow junction cells under 30 suns concentration[9]. Recombination in crystalline silicon cells for the low injection regime and for V > 0.3 volts is often dominated by bulk recombination in the QNR base.

3.5 CURRENT-VOLTAGE CURVE AND EFFICIENCY

The current-voltage (I-V) curve for a solar cell is obtained by plotting the output current as a function of the output voltage. This is done by varying the load from zero ohms to several kilo-ohms while the cell is illuminated (fig. 3.5-1). Alternatively, when

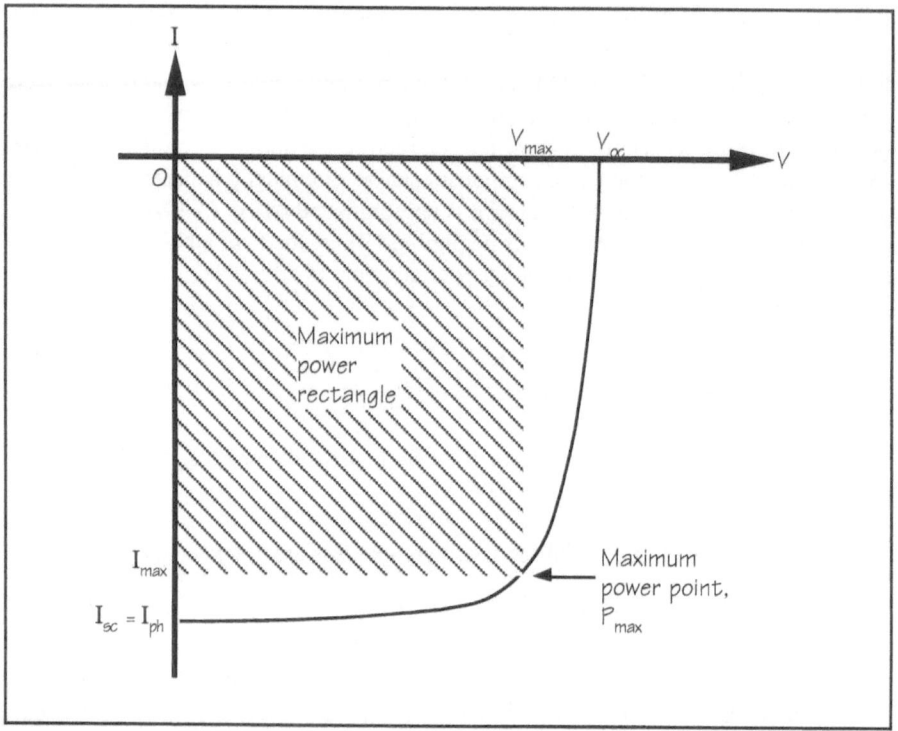

Fig. 3.5-1   Current vs. voltage for an illuminated pn-junction solar cell. The series
resistance is assumed to be very small and the shunt resistance very large, so that $I_{sc} = I_{ph}$.
The load R_L is adjusted so that the cell operates at P_{max}.

the shifting approximation is valid, one can simply measure the dark I-V characteristic
and shift it by the short circuit current (fig. 3.5-2). The illuminated curve is in the fourth
quadrant of the I-V plane indicating that power is being supplied to the load. For convenience, the absolute value of the current can be used. This reflects the curve about
the x-axis and puts it in the first quadrant (fig. 3.5-3) – the configuration shown in
manufacturers' data sheets.

The solar cell delivers maximum power to the load when the load is adjusted so that the
product I x V is maximum. This corresponds to the point on the curve labeled (I_{max},
V_{max}) in fig. 3.5-1. In any practical application, the load is adjusted so the cell (or
module) operates at or near this point on the curve. Efficiency η is the ratio of output

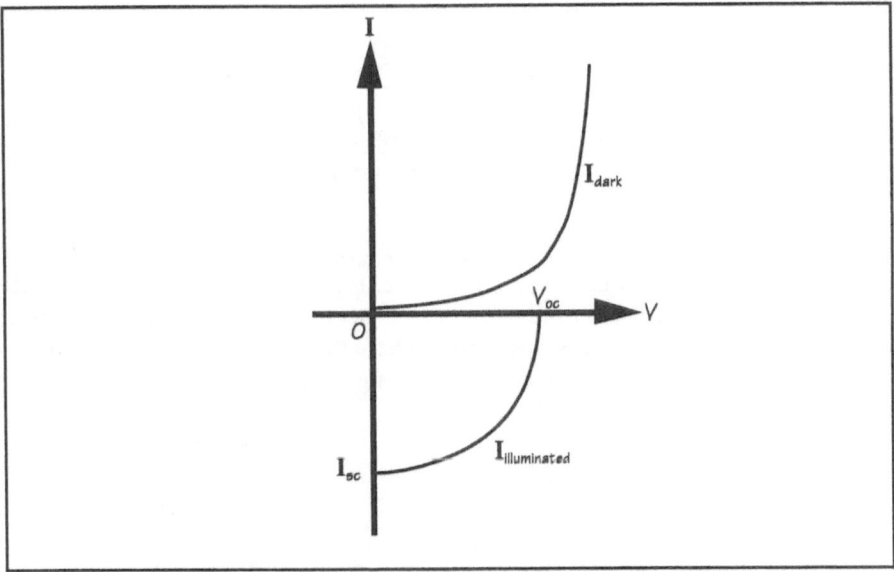

Fig. 3.5-2   With the shifting approximation, illuminated current is the dark current shifted by I_{sc}.

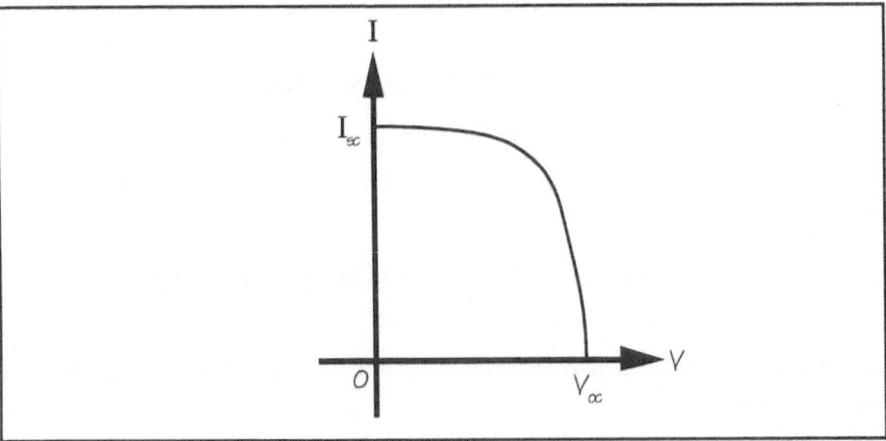

Fig. 3.5-3   I-V curve reflected into the first quadrant.

power to input power:

$$\eta = P_{output}/P_{input}$$

$$= I_{max}V_{max}/P_{input}$$

$$= I_{sc}V_{oc}FF/P_{input} \qquad (3.5\text{-}1)$$

where the factor FF is defined as the ratio $I_{max}V_{max}/I_{sc}V_{oc}$. FF is called the *fill factor*, and is a measure of the squareness of the curve. Equation (3.5-1) indicates the efficiency of a solar cell is maximized when the product of the short circuit current, open circuit voltage, and fill factor is maximized. These factors are coupled to each other, and the maximization of one factor through a design or processing change often leads to a degradation of the other two factors.

A. Short Circuit Current

As light passes into a cell, the photon flux per unit bandwidth ($cm^{-3} s^{-1}$) at a depth x is

$$F(x, \lambda) = F(0, \lambda)e^{-\alpha x}$$

where $F(0, \lambda)$ is the flux passing into the top surface, and the absorption coefficient α is a function of wavelength λ. The pair generation rate per unit bandwidth at depth x is

$$-\frac{d}{dx}F(x, \lambda) = F(0, \lambda)\alpha e^{-\alpha x}$$

and has units of $cm^{-4}s^{-1}$. The pair generation rate for the entire cell is then

$$G = A \int_0^W \int_0^{hc/E_g} F(0, \lambda)\alpha e^{-\alpha x} d\lambda dx$$

where hc/E_g is the wavelength corresponding to the bandgap, A is the area, and W is the width of the cell.

Short circuit current is determined by the cell's quantum efficiency – the fraction of incident photons that produce electron-hole pairs at a given wavelength. *Internal* quantum efficiency is the fraction of photons transmitted into the material that result in pairs

$$IQE(\lambda) = (G/A)(hc/\lambda P_{in}) \approx (J_{sc}/q)(hc/\lambda P_{in}) \qquad (3.5\text{-}2)$$

where P_{in} is the input power density transmitted across the surface at wavelength λ (fig. 3.5-4). *External* quantum efficiency is the fraction of incident photons to pairs

$$EQE(\lambda) = IQE(\lambda) \, [1\text{-}R(\lambda)]$$

where $R(\lambda)$ is the reflection coefficient at λ. The photocurrent per unit bandwidth at

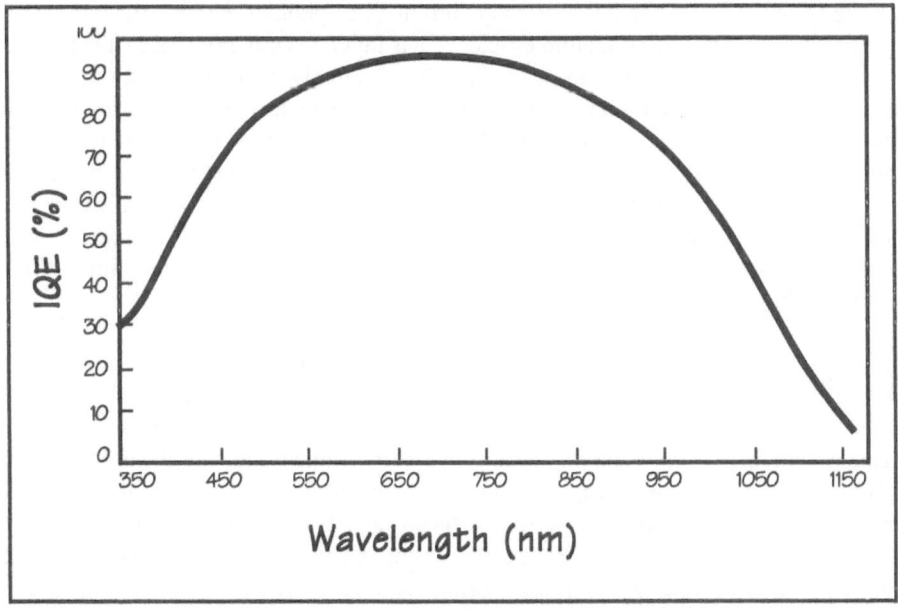

Fig. 3.5-4   Representative spectral response curve for float-zone silicon cell.

wavelength λ is then the external quantum efficiency times the spectral distribution at λ

$$I_{ph}(\lambda) = qAS(\lambda) \, EQE(\lambda) \qquad (3.5\text{-}3)$$

where A is the area of the cell and $S(\lambda)$ is the spectral distribution or number of incident photons per unit time per unit bandwidth per unit area at λ. The integral of eq. (3.5-3) over all wavelengths yields the total photocurrent

$$I_{ph} = \int_0^\infty qAS(\lambda) \; EQE(\lambda) \; d\lambda \qquad (3.5\text{-}4)$$

Only a limited part of the solar spectrum will produce useable, i.e., collectable, electron-hole pairs. At the ultraviolet end of the spectrum, the skin depth is so short that all of the light is absorbed in the emitter before reaching the junction. The minority carriers that are produced by these short wavelengths will recombine in the heavily-doped emitter almost immediately. On the other hand, for wavelengths longer than the bandgap ($\lambda = 1.1$ μm), the skin depth is effectively infinity, i.e., photons are insufficiently energetic. By $\lambda = 1.2$ μm, the quantum efficiency has fallen to zero. For those wavelengths that can be absorbed in the base (wavelengths of roughly 0.3 to 1.1 μm), the magnitudes of the diffusion length L, cell width W, and skin depth $1/\alpha$ (λ) are critical to the internal quantum efficiency. If $L < 1/\alpha$ (λ) $< W$, the light will be absorbed in the base, but few of the photogenerated minority carriers will reach the junction. An example is red light (0.7 μm) in a 300-μm cell for which $L = 150$ μm. The value of $1/\alpha$ at this wavelength is about 200 μm, and most of the photogenerated minority carriers will recombine in the base when they are still about 50 μm from the junction. Short diffusion lengths cause a fall off of the IQE curve at long wavelengths.

The short circuit current is improved by using a shallow junction depth and a high base resistivity. A shallow junction depth improves I_{sc} by increasing the fraction of short wavelength (ultraviolet) photons that can reach the base before being absorbed. This is particularly important for space cells where a large fraction of the light is in the blue end of the spectrum. A high base resistivity means that the ionized dopant density is low, and thus, the minority carrier diffusion length is improved. More minority carriers can reach the junction. For single-crystal and polycrystalline silicon cells, the junction depth is typically 0.4 to 0.6 μm; the base resistivity is about 1 Ω cm, which corresponds to a boron doping density of 1.5 x 10^{16}/cm³.

The polarity of the emitter and base (n⁺ on p, or p⁺ on n) also affects the short circuit current. Silicon cells are fabricated to be *n⁺ on p* to take advantage of the larger electron diffusion length in the base of an n⁺/p cell, as opposed to the hole diffusion length in the base of a p⁺/n cell, for similar base resistivities.

The short circuit current is degraded by series resistance. Any series resistance will bias the dark diode, and cause the current through the load (short circuit) to be less than

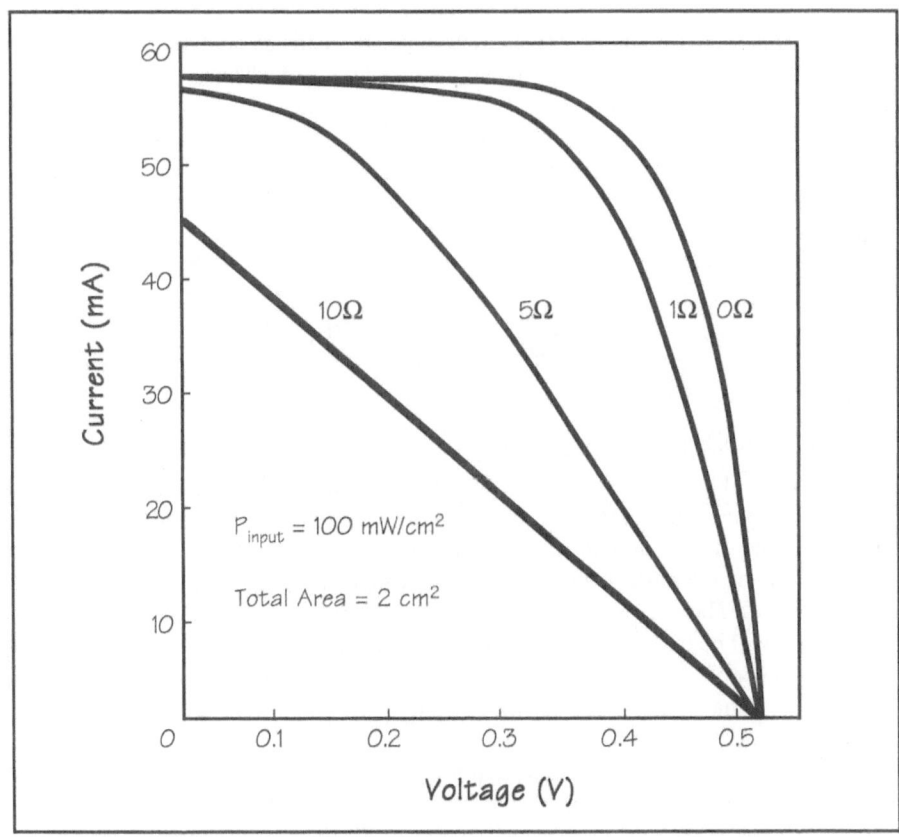

Fig. 3.5-5   The effect of series resistance on the fill factor and short circuit current
of a silicon solar cell. (*After H.J. Hovel, Semiconductors and Semimetals, vol. 11, Academic
Press, 1975.*)

the photogenerated current. This effect is seen in fig. 3.5-5. The reduction in J_{sc} by the
series resistance becomes very pronounced when the fill factor is less than about 0.6.

B. Open Circuit Voltage

The open circuit voltage is a measure of how strongly the separated photogenerated
charge can bias the junction for any given level of illumination. Substitution of

eq. (3.4-3) into eq. (3.4-1), with the circuit opened, yields an analytical expression for V_{OC}. When the circuit is opened, the output current is zero. Thus,

$$0 = I_{SC} - I_D$$
$$= I_{SC} - I_o \left(e^{qV/nkT} - 1 \right)$$

and

$$V_{OC} = \frac{nkT}{q} \ln \left[\frac{I_{SC}}{I_o} + 1 \right] \tag{3.5-5}$$

where ln refers to the natural logarithm. Equation (3.5-5) shows the open circuit voltage of the cell increases as the saturation current I_o decreases. As indicated in section 3.4, the saturation current is determined by the carrier recombination in both the QNRs and the junction SCR. For both regions, the rate of carrier recombination[10,11] is dependent on the intrinsic concentration n_i, which is a temperature-dependent material parameter. The greater is n_i, the greater will be the recombination rates, and thus, the greater will be the saturation current I_o. Since large n_i is characteristic of a narrow-bandgap material, it is correctly expected that silicon cells ($E_g = 1.12$ eV) have smaller values of V_{OC} than GaAs cells ($E_g = 1.42$ eV).

For the *ideal diode* case, an increase in n requires an increase in I_o. If n is 1, I_o only consists of recombination in the QNRs. But if n is greater than 1, then I_o also includes a recombination component from the junction SCR. For the ideal case represented by eq. (3.5-5), maximization of V_{OC} requires the dark current to be dominated by recombination in the QNRs and n = 1. Experimentally, this sometimes proves *not* to be the case[3,12]. For the world record efficiency silicon cell discussed in[3] and[12], the I-V curve (see fig. 3 in[3]) has a pronounced component with n ≈ 2, and a corresponding huge saturation current density (about 10^{-9} A/cm²) for the diode. Yet V_{OC} is over 700 mV! The observation that there is little correlation between values of n and V_{OC} has also been seen in some GaAs and CdTe cells[13]. Again, this demonstrates that non-idealities are common, and supports the unorthodox concept that recombination is driven by transport mechanisms, i.e., the I-V curve, rather than the other way around. Thus, at least in some cases, it appears that values of n and J_o are correlated with recombination, but are not determined by recombination. For these cases, device optimaization seeks methods of shifting the dark I-V curve to the right by manipulating the junction rather than methods of lowering recombination rates[14].

Open circuit voltage is degraded by shunt resistance. When the circuit is opened at the load, the current path through the shunt resistance lowers the bias across the diode thereby degrading V_{oc}. This effect is seen in fig. 3.5-6.

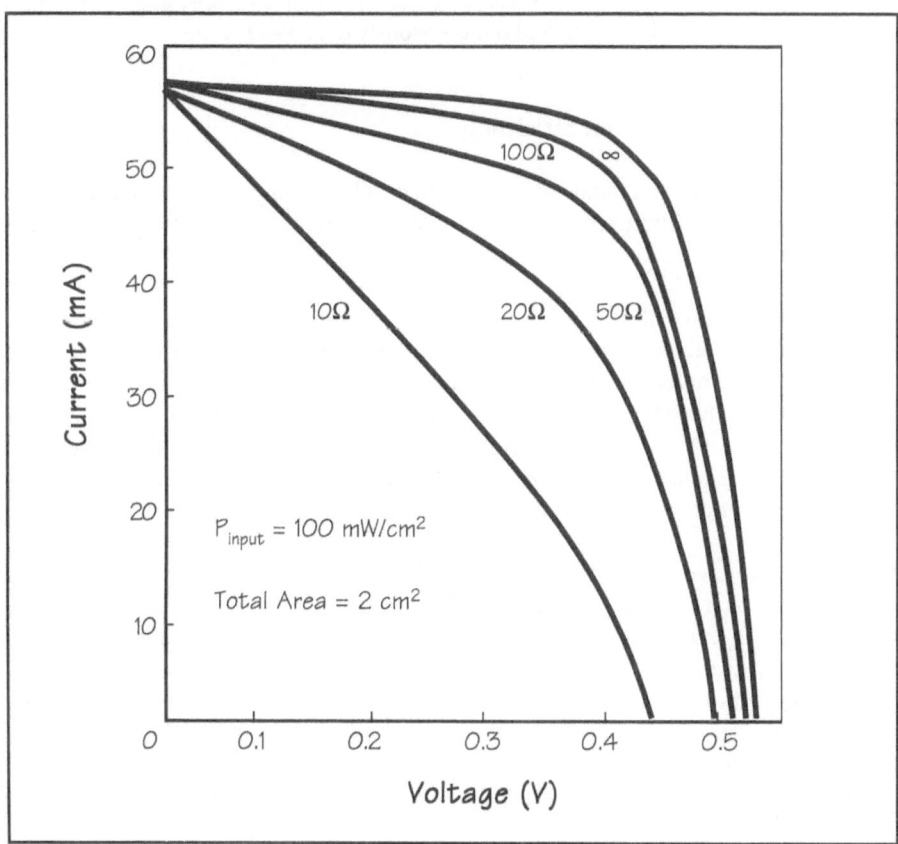

Fig. 3.5-6 The effect of shunt resistance on the fill factor and open circuit voltage of a silicon solar cell. (*After H.J. Hovel, Semiconductors and Semimetals, vol. 11, Academic Press, 1975.*)

C. Fill Factor

Fill factor reflects the effects of series and shunt resistances on the I-V characteristic. Series resistance degrades the fill factor by increasing the bias across the diode. Since the diode current is exponentially dependent on the diode bias, the parasitic diode current is greatly increased by even a small series resistance. Shunt resistance degrades the fill factor by diverting some of the photogenerated current that would otherwise go through the load R_L. In either case, the knee of the curve becomes more rounded and the output power decreases. For a crystalline silicon cell, series resistance as large as a few ohms or shunt resistance as small as a hundred ohms seriously degrades I-V curve squareness.

D. Effect of Temperature on Efficiency

Efficiency is a function of temperature through the effects on short circuit current, open circuit voltage, and fill factor. This is particularly important for cells that operate in temperature extremes.

In crystalline silicon cells, the short circuit current improves slightly with temperature over a wide temperature range because of improvement in the minority carrier diffusion length. As temperature increases, the thermal velocity of carriers increases, and it becomes less probable that excess minority carriers in the base will recombine with the surrounding majority carriers. As a result, the percentage of excess minority carriers that can reach the junction and be transferred to the emitter will increase with temperature. Thus, the photogenerated current increases with temperature.

Open circuit voltage and fill factor are degraded by temperature. Degradation of open circuit voltage is not expected from a quick glance of eq. (3.5-5). However, the saturation current I_o is an increasing function of the intrinsic carrier concentration n_i, and n_i is exponentially dependent on the temperature. This exponential dependence, coupled with the logarithm in eq. (3.5-5), causes V_{OC} to decrease almost linearly with temperature. The fill factor is degraded with rising temperature partly because of the increase in series resistance. The effect of rising temperature on I_{SC}, V_{OC}, and fill factor is illustrated by the measured data in fig. 3.5-7.

The net effect of rising temperature is to decrease efficiency in most cases. Because of second-order semiconductor effects, this is sometimes not true at very low temperatures (less than -50°C) as might be encountered in some space missions when the cell is obliquely illuminated.

E. Limits to Solar Cell Conversion Efficiency

For any material system, there are intrinsic limitations to photovoltaic energy conver-

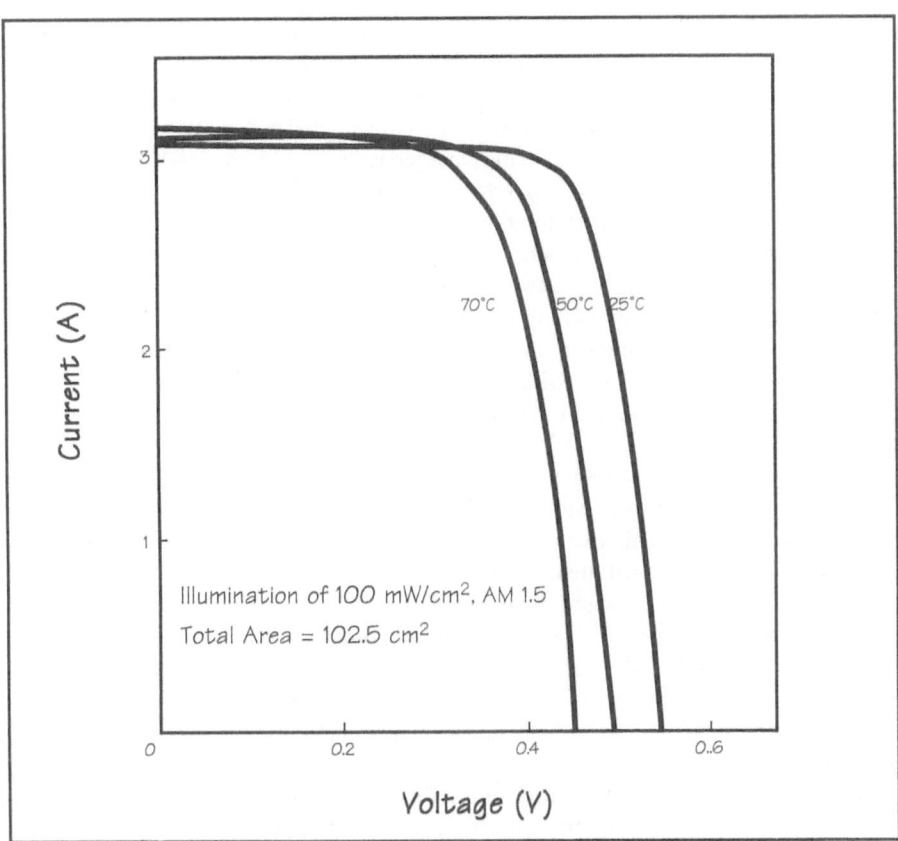

Fig. 3.5-7   Effect of temperature on I_{sc}, V_{oc}, and fill factor for a semicrystalline silicon cell. Efficiency at 25° C is 12.7%. (*Data courtesy Photocomm, Inc.*)

sion. These include open-circuit voltage and fill factor limitations. The value of V_{oc} can never exceed the built-in voltage. If it did, the built-in field would be cancelled by the excess majority carriers on the two sides of the junction, and excess minority carriers in the quasi-neutral regions could no longer be separated. With regards to the fill factor, by referring to the cell I-V characteristic, it is seen that only the area in the rectangle bounded by V_{oc} and I_{sc} contributes to the power delivered to the load. The remaining power between the curve and the rectangle is lost through series and shunt resistances. The ideal cell would include zero series resistance and infinite shunt resistance in the equivalent-circuit model. Next, there is the limitation imposed by the mismatch between the magnitude of the bandgap and the energy of incoming photons. As previously mentioned, photons whose energy is less than the bandgap are not absorbed. The material is transparent. Finally, electron-hole pairs created by photons much more energetic than the bandgap recombine through radiative and non-radiative processes so that much of the excess energy above the bandgap is lost, i.e, thermalization. The effect of spectral mismatch to the bandgap is ameliorated by using successive junctions with the bandgap decreasing from the top down. This is the approach in a tandem junction cell. The first junction is well-matched to high energy photons, the second junction to slightly less energetic photons, and so on. The junctions are connected by very heavily doped interfaces that allow carriers to quantum mechanically tunnel through to the other side. In theory and practice, this approach increases conversion efficiency compared to a single junction. Cells with more than two tandem junctions are possible but not very feasible from a fabrication standpoint.

With the sun and cell modeled as blackbodies (i.e., perfect absorbers/radiators), the conversion efficiency is shown to be a function of the bandgap and the ratio of the cell and sun temperatures. This treatment predicts a maximum possible conversion efficiency of about 40% at a cell temperature of 300 K[15]. Though the modeling of the cell as a blackbody is a rough approximation, it allows the prediction of upper limits to efficiency. Other treatments[16, 17] based on thermodynamic considerations also model the cell as a blackbody and predict maximum conversion efficiency values of between 30% and 37%, respectively, for single-junction cells. Lower values are predicted by semi-empirical studies[18, 19] which include the fundamental limitations listed above. In general, efficiency curves peak as a function of bandgap and air mass. Studies are consistent in their prediction of maximum efficiency at a bandgap value in the 1.35 to 1.45 eV range for AM0 conditions. Cells with more than two tandem junctions are possible and n^+p GaInP$_2$/n^+p GaAs/Ge-substrate space cells will evolve in this direction for efficiencies approaching 35%.

3.6 RADIATION EFFECTS

Long-term power generation on space vehicles is provided by photovoltaics, providing the vehicle remains sufficiently close to the sun. Performance is degraded by cumulative radiation damage from the broad spectrum of electrons and protons emitted by the sun's fusion processes. While the flux of energetic particles at the earth's surface is negligibly small, fluxes outside the atmosphere are large enough to cause much degradation to cell performance over periods of months or years. The particle radiation environment in the vicinity of the earth consists mostly of electrons and protons trapped for long time intervals in the geomagnetic field. This plasma is distorted by the solar wind, a plasma of mostly protons with average energy of 1 keV and a density of only about 10 cm^{-3}. Particles associated with solar flares have little effect on cells. Galactic cosmic ray radiation, while very energetic (1 GeV), has negligible effect because of its small flux. As charged particles, electrons and protons are accelerated in a helical fashion along the earth's magnetic field lines.

In general, low-earth orbits have relatively benign radiation environments. At 500 km altitude (28°inclination), trapped 1-MeV electron and 10-MeV proton fluxes are both about 6 x 10^5 particles/cm^2/day[20]. At the altitude of geosynchronous orbit (35,800 km), the trapped 1-MeV electron flux is 5 X 10^{10} particles/cm^2/day, while the trapped proton flux is negligible. Electron energies at geosynchronous altitude reach several MeV, with the vast majority of electrons having energies below 2 MeV[21]. In geosynchronous orbit, dose rate is longitudinally dependent because the geomagnetic and geographic coordinate systems do not coincide[22]. Common radiation test regimes for characterizing total dose hardness are 1-MeV electrons and 10-MeV protons. An array in geosynchronous orbit receives a total dose of 10^{15} cm^{-2} 1-MeV electrons in an exposure time of 1700 to 4000 days. For non-geosynchronous orbit, dose rate variations at different longitudes are averaged out during the day so that a given total dose corresponds to a specific time interval. It is the intermediate altitudes that have high fluxes of very energetic particles. This includes a high flux of 10-MeV protons in the vicinity of 1.5 earth radii. Until recently, these altitudes were avoided because of the poor end-of-life (EOL) performance of cells. Total dose rates from trapped particles for 500 km (28° inclination) and 700 km (sun synchronous) orbits are 3 x 10^5 and 4 x 10^7 rads (Si) per year, respectively. At 1.5 earth radii, a total dose of 10^{13} cm^{-2} 10-MeV protons is expected with 580 days in orbit[23].

For solar cells, the major radiation damage mechanisms are lattice displacement, and to a lesser extent, ionization. Degradation from both mechanisms is a function of total dose. Lattice displacement is the process whereby an elestic collison between the nucleus of a regularly placed atom and an energetic particle causes the atom to move into an interstitial position. The resulting stable vacancy-interstitial pair easily migrates (with an activation energy characteristic of the semiconductor) and forms defect complexes with other defects or with impurities. This results in the creation of recombination sites (traps) near the middle of the bandgap and reduction of minority carrier diffusion length. Several atoms can be displaced from their lattice sites by a single particle; divancies are a major defect in heavily irradiated material[24]. The equilibrium majority carrier concentration also changes because of interaction of the vacancy with oxygen and donor atoms. In n-type Si, vacancies react with oxygen to form vacancy-oxygen (V-O) pairs, and with donor atoms (phosphorus) to form vacancy-donor (V-P) pairs. Both defects are electrically active. By accepting an electron from the conduction band, they become negatively charged and lower the equilibrium majority electron density. The V-O and V-P pair defects reside at 0.17 eV and 0.4 eV below the bottom of the conduction band, respectively. To a lesser extent, a similar phenomenon occurs in p-type silicon. Energetic electron irradiation creates vacancy-boron pairs that reside about 0.3 eV above the top of the valence band. The defect behaves like a donor and accepts a hole from the valence band. This decreases the equilibrium majority hole concentration. Since the conductivity of the material is directly proportional to the majority carrier density, majority carrier loss increases the series resistance. This degrades the fill factor. However, I_{sc} and V_{oc} often suffer more than FF because the mobility of carriers (majority carriers) is dominated by the density of ionized dopant atoms except when there is a very high concentration of vacancy-interstitial defects. Minority carrier loss degrades both I_{sc} and V_{oc}. For most material systems, I_{sc} is degraded more severely than V_{oc}.

Lattice displacement damage in silicon can be repaired by thermal annealing if sufficiently high temperatures can be achieved. In the annealing process, the displaced atoms return to their original positions and the defect disappears. Vacancy-phosphorus pairs and V-O pairs anneal rapidly at about 150° C and 350° C, respectively[25]. Systems that develop these temperatures are not feasible in space deployment. However, cells based on other material systems, e.g., InP, undergo self-annealing at lower temperatures.

Ionization is the inelastic collison process whereby an atomic electron gains enough energy from a collision to become unbounded. An electron moves from the valence to the conduction band and creates a hole-electron pair. This is similar to the generation of a pair by an incoming photon. Ionization by collision requires about three times the energy as ionization by absorbing a photon. Ionization degradation in silicon cells occurs as electrons becomes trapped in the surface oxide or other dielectric. This increases leakage current at the surface of the cell and degrades the fill factor.

A figure of merit for the radiation hardness of a particular material is the diffusion length damage coefficient K_L, defined by

$$(1/L)^2 = (1/L_O)^2 + K_L \phi, \qquad (3.6\text{-}1)$$

where L and L_O are the post- and pre-irradiation diffusion lengths, and ϕ is the radiation fluence (particles/cm^2). If L $<<$ L_O, then $K_L = 1/L^2\phi$, and $\log L = -0.5 (\log \phi) - 0.5 (\log K_L)$.

For sufficiently high fluence, the diffusion length is degraded, and the plot of log L vs. log φ is a straight line. The intercept and the value for K_L then define the particle type and energy dependence of the cell's degradation. This is illustrated in fig. 3.5-8 for an n⁺p silicon cell[26]. To ameliorate particle and UV radaition damage, 5% cerium oxide glass covers are fixed to cells with adhesive[27].

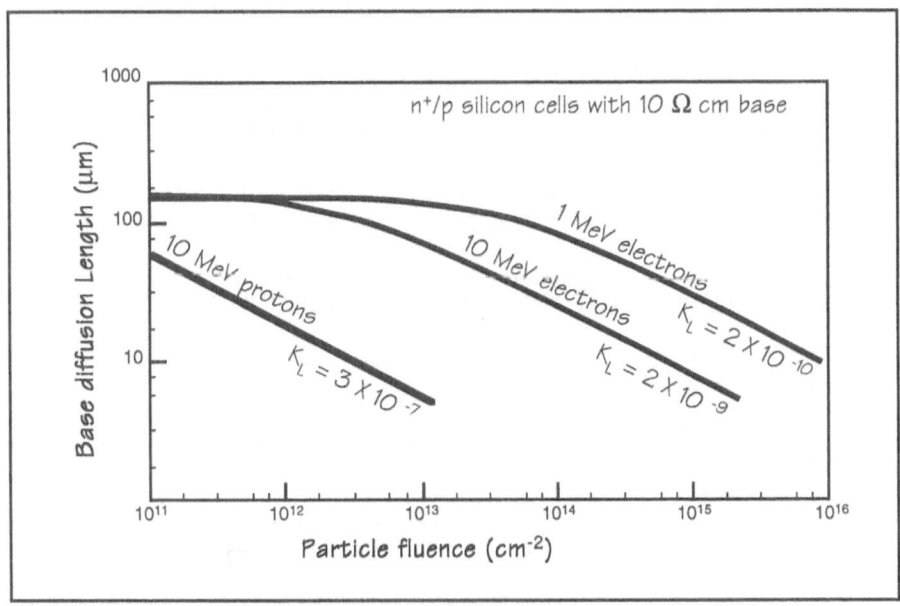

Fig. 3.5-8   Degradation of diffusion length with fluence for various radiations. (After H.Y. Tada, et al., *Solar Cell Radiation Handbook*, 3rd ed., JPL publication 82-69, Feb. 1989.)

REFERENCES

[1]   B. von Roedern, "Innovative optimization procedures for solar cells based on a unique model for junction optimization," *Proc. of the 12th European Photovoltaic Solar Energy Conf.*, pp. 1354-1358, April 1994.

[2]   B. von Roedern, "Improved solar cell processing concepts from the use of fuzzy logic principles," *Proc. of the 13th European Photovoltaic Solar Energy Conf.*, pp. 154-157, Oct. 1995.

[3]   P. Altermatt, et al., "Analysis and minimization of resistive losses in high-efficiency Si solar cells by combining measurements with numerical modeling," *Proc. of the 13th European Photovoltaic Solar Energy Conf.*, pp. 382-385, Oct. 1995.

4   J.A. Mazer, A. Neugroschel, and F.A. Lindholm, " A method for experimental assessment of the shift-
        ing approximation, with application to polysilicon solar cells," *IEEE Trans on Electron Devices*, vol.
        ED-28, pp. 1530-1534, Dec. 1981.

5   F.A. Lindholm, J.G. Fossom, and E.L. Burgess, "Application of the superposition principle to solar cell
        analysis," *IEEE Trans. on Electron Devices*, vol. ED-26, pp. 165-171, March 1979.

6   N.G. Tarr and D.L. Pulfrey, "The superposition principle for homojunction solar cells," *IEEE Trans. on
        Electron Devices*, vol. ED-27, pp. 771-776, April 1980.

7   C.T. Sah, R.N. Noyce, and W. Shockley, "Carrier generation and recombination in p-n junctions and p-
        n junction characteristics," *Proc. IRE*, vol. 45, pp. 1228-1243, Sept. 1957.

8   A. Neugroschel, F.A. Lindholm, and C.T. Sah, "A method for determining the emitter and base life-
        times in p-n junction diodes, " *IEEE Trans. on Electron Devices*, vol. ED-24, pp. 662-671, June 1977.

9   B. Affour, et al., "Relation of recombination mechanisms with junction, bulk, and back surface energy losses in
        solar cells," *Proc. of the 12th E.C. Photovoltaic Solar Energy Conf.*, pp. 1347-1349, April 1994.

10   C.T. Sah, R.N. Noyce, and W. Schockley, "Carriers generation and recombination in p-n junctions and
        p-n junction characteristics," *Proc. of the IRE*, vol. 45, pp. 1228-1243, Sept. 1957.

11   W. Shockley and W.T. Read, "Statistics of the recombination of holes and electrons," *Phys. Rev.*, vol.
        87, pp. 835-842, Sept. 1952.

12   J. Zhao, "24% efficient silicon solar cells," *Conf. Rec. of the First World Conf. on Photovoltaic Energy
        Conversion*, pp. 1477-1480, Dec. 1994.

13   B. von Roedern, "Higher Efficiencies through defect engineering of solar cell junctions," *11th
        E.C.Photovoltaic Solar Energy Conf.*, pp. 295-298, Oct. 1992.

14   K.A. Bertness, et al., in *AIP Conference Proceedings No. 306*, p. 106, American Institute of Physics,
        Woodbury, NY, 1994.

15   A.M. Buoncristiani, C.E. Byvik, and B.T. Smith, "Thermodynamic limits to the conversion of black-
        body radiation by quantum systems," *J. Appl. Phys.*, vol. 53, pp. 5382-5386, Aug. 1982.

16   C.H. Henry, "Limiting efficiencies of ideal single and multiple energy gap terrestrial solar cells," *J.
        Appl. Phys.*, vol. 51, pp. 4494-4500, Aug. 1980

17   W. Shockley and H.J. Queisser, "Detailed balance limit of efficiency of p-n junction solar cells," *J.
        Appl. Phys.*, vol. 32, pp. 510-519, Mar. 1961.

18   J.J. Loferski, "Theoretical considerations governing the choice of optimum semiconductor for photo-
        voltaic solar energy conversion," *J. Appl. Phys.*, vol. 27, pp. 777-784, July 1956.

[19]   D.H. Hartman and C.C. Shen, "GaAs photodetectors," in *Gallium Arsenide Technology*, ed. by D.K. Ferry, pp. 384-386, Howard W. Sams & Co., Indianapolis, 1985.

[20]   *TRW Space Data Book*, 4th edition, ed. by N.J. Barter, pp. 2-3 to 2-5, TRW Space and Technology Group, Rodondo Beach, CA, 1996.

[21]   H.Y. Tada, *et al.*, *Solar Cell Radiation Handbook*, 3rd ed., section 5.1, Jet Propulsion Laboratory publication 82-69, Nov. 1982.

[22]   E.G. Stassinopoulos, "The geostationary radiation environment," *J. Spacecraft and Rockets*, vol. 17, pp. 145-152, (1980).

[23]   D.M. Sawyer and J.I. Vette, *AP8 Trapped Proton Environment for Solar Maximum and Solar Minimum*, NASA document 77N-18983, p. 93, Dec. 1976.

[24]   D.P. Parton and T. Markvart, "Recombination centres in solar cells: DLTS study," *Proc. of the 12th E.C. Photovoltaic Solar Energy Conf.*, pp. 508-510, Apr. 1994.

[25]   Tada, *et al.*, *Solar Cell Radiation Handbook*, 3rd ed., p. 3-11, Jet Propulsion Laboratory publication 82-69, Nov. 1982.

[26]   Tada, *et al.*, *Solar Cell Radiation Handbook*, 3rd ed., p. 3-22, Jet Propulsion Laboratory publication 82-69, Nov. 1982.

[27]   Optical Coating Laboratory, Inc., Solar Products data sheet for radiation resistant glass covers, Santa Rosa, CA, May 1996.

4

CELL AND MODULE DEVELOPMENT

From the standpoint of production costs and division of labor, there are four main tasks in crystalline silicon PV manufacturing: growth of the semiconductor material, separation into wafers, junction formation and cell processing, encapsulation and framing. In 1995, worldwide module shipments were still dominated by non-ribbon (i.e., ingot-based) crystalline silicon technology. These modules required all four processing areas for their manufacture. Ribbon and thick film deposition technologies make the second task less expensive. To relieve the expense of the fourth task, frameless modules using adhesive bonding to a support structure are being developed as an alternative to an integral frame. A potentially inexpensive approach is a ribbon or thick film in a frameless module. Much work in the late 1990s will be done to exploit these cost reducing innovations.

The smallest deployable photovoltaic unit is the module. Modules are designed small enough to be conveniently handled by installation personnel. They can stand alone or be electrically connected to form a larger power unit, i.e., an array. A module contains one or more strings of series-connected cells. This renders a useful voltage. When cells are connected in series, their voltages add. However, when connected in parallel, the currents add only if the individual currents are equal. For the case of two unequal cell currents, the larger current is partly bled through the diode and shunt resistance of the cell with the smaller current. A common configuration is a module built from thirty ingot-based semicrystalline silicon cells that each measure 10 cm x 10 cm. The open-circuit voltage, short-circuit current density, and fill factor of the cells are approximately 600 mV, 31 mA/cm^2, and 0.77, respectively, under standard test conditions. This yields a cell efficiency of 14%. One possible configuration is a 30-unit string. The open circuit voltage of the module then approaches 18 V, and the short-circuit current approaches 3.2 A. If the fill factor of the module is 0.75, the corresponding voltage and current at the maximum power point are about 14 V and 2.8 A, respectively. This is a good configuration for charging a nominal 12-volt storage battery. Alternatively, the module can be configured with two parallel equal-current strings of 15 cells each. At the maximum power point, the module's output voltage and current are then about 7 V and 5.6 A, respectively – a good configuration for charging 6-volt batteries. Often module circuitry is flexibly designed so the user can switch between the one-string or two-string modes simply by an easy rearrangement of connections in the junction box. A complete photovoltaic charging system includes a charge controller so that batteries are not overchargerd, and an inverter if the system is connected to the utility grid (fig. 4.0-1). Efficiencies for small laboratory-scale single-crystal silicon modules (862 cm^2) have reached 21.6% AM1.5 spectrum (100 mW/cm^2) at 25°C[1]. Commercial semicrystalline and single-crystal modules are in the 12% to 14% range, with areas of about 0.5 to 2 m^2.

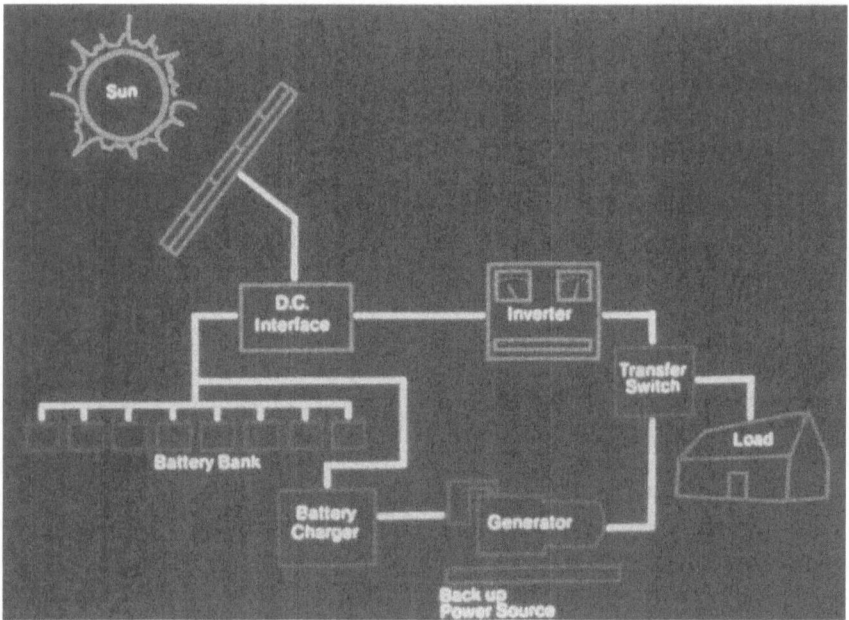

Fig.4.0-1   Schematic of stand-alone PV system. (*Photo courtesy SEIA*)

4.1 PROCESS FLOW

Table 2 shows the.process flow for cell and module manufacturing.

Considerable variation exists between commercial cell fabrication processes because of inherent differences in technologies, exclusivity of patented processes, and differ-

PROCESS FLOW FOR COMMERCIAL TERRESTRIAL CELL AND MODULE MANUFACTURING

Growth of Semiconductor

Czochralski or float-zone ingot
Polysilicon cast ingot
Ribbon or thick film on ceramic substrate

Separation into Wafers

Internal diameter saw or wire saw for ingots
Wet etchant NaOH solution to remove surface damage
Diamond or laser scribing for ribbons and thick films

Cell Fabrication

Surface cleaning with H_2SO_4/H_2O_2 or $NH_4OH/H_2O_2/H_2O$
Surface texturing of front side with KOH or NaOH solution (*single crystal only*)
Phosphorus diffusion in tube or belt furnace
Remove phosphosilicate glass with 10% HF solution (*surface hydrophobic*)
Remove n^+ dopant on the edges
Passivation of surface and bulk with plasma enhanced chemical vapor
   deposition (PECVD) SiN_x film, or surface with thermal oxide
Aluminum paste and alloy on rear for contact, back-surface field, and gettering
Silver paste contact pads on backside
Screen print silver paste front contacts
Fire contacts simultaneously – front paste fired through passivation /
   anti-reflection coating
Forming gas (10% H_2/90% N_2) anneal
Anti-reflective coating by TiO_2 CVD or other AR coating

Encapsulation and Framing

Sort cells according to short-circuit current
Attach solder-plated copper leads
Layout matched cells with front-to-back ribbon connections
Form 5-layer sandwich: glass, EVA, cells, EVA, Tedlar backplane
Place laminated sandwich in rigid (*metal*) frame
Attach junction box
Record I-V characteristic of module

Table 2   Process flow for cell and module manufacturing

ences in efficiency goals. The most significant variations include the following.

(1) Some processes grow a thermal SiO_x layer at the end of the phosphorus diffusion as a passivant to lower surface recombination velocity. The phosphosilicate glass is not removed. Alternatively, at the end of the phosphorus diffusion, SiN_x is deposited by plasma enhanced chemical vapor deposition (PECVD) between 300 and 400°C after de-glazing. Deposition of silicon nitride serves both as an anti-reflection coating and a passivant.

(2) Some semicrystalline processes do not use a back-surface field to improve open-circuit voltage because the extra step does not greatly affect cost per watt.

(3) Some spray-dopant processes obviate edge isolation by employing capillary absorption of the liquid dopant source near the edges of the wafer.

(4) Some processes deposit the anti-reflection coating after the metallization by evaporating a TiO_2 and/or Al_2O_3 film. In another method, a highly uniform TiO_2 film is formed by low-temperature (325°C) pyrolysis of a titanium-containing spin-on glass after the metallization[2].

(5) Processes using cast material seldom use backside gettering because of the minimal benefit when the minority carrier diffusion length is only about half the width of the cell

(6) High-efficiency cells use evaporated metallization (Ti/Pd/Ag) instead of screen printing for the front side contacts. This provides minimal interfacial contact resistance and avoids the blue-response degradation associated with deep junctions and highly doped surface layers necessary for screen printing. The disadvantage is the expense of a slower process.

(7) An emitter etch-back between the contact fingers is used in high efficiency cell processes so that blue response is enhanced while maintaining low series resistance under the fingers.

4.2 PN-JUNCTION FORMATION

Emitters in commercial silicon cells are formed by phosphorus diffusion. The concentration profile more or less follows the solution of a classical one-dimensional *diffusion equation* except very near the surface where the dopant can pile up above the solid solubility limit. This equation is similar to that which governs the transport of heat through a homogeneous solid. Validity of the diffusion equation is dependent on the assumptions that the dopant atoms obey classical mechanics and have many fewer collisions with each other than with the lattice. The exact solution is determined by the boundary and initial conditions. In semiconductor technology, there are two important sets of boundary and initial conditions for dopant diffusions into the bulk. These two diffusion regimes are known as *predeposition* and *drive-in*. Both assume that the bulk is semi-infinite, i.e., the thickness of the bulk is much greater than the mean diffusion distance of the dopant. In a predeposition, the surface concentration C_S of impurity is held constant. In a drive-in, the total quantity Q of impurity in the bulk is held constant.

The n^+/p polarity is chosen over the p^+/n polarity because of a slight efficiency advantage that exists over a wide range of base resistivities. For any given base resistivity, electron diffusion length is greater than hole diffusion length. This increases both short-circuit current density and open-circuit voltage (due to lower J_o) in n^+/p cells compared to p^+/n cells with equal base resistivity. The diffusion length in the base of p^+n cells can be increased by lowering the base doping. But then the built-in voltage (which limits the open-circuit voltage) is degraded. For a wide range of base resistivities, the $J_{sc}V_{OC}$ product is larger in the n^+/p polarity configuration. An additional (and perhaps more important) reason for choosing the n^+p configuration is that phosphorus diffusions are easier to control than boron diffusions. This is related to non-idealities in the diffusion of any impurity, including phosphorus, from a glassy layer into the bulk[3]. For the short diffusions characteristic of emitter formation, boron is more problematic in this regard than phosphorus.

Base doping concentration is uniform because the acceptor impurity (boron) has an impurity distribution coefficient close to unity (0.8). The boron is well-mixed with the molten silicon by convection before growth begins. The emitter doping concentration, though, has a large gradient. At the metallurgical junction, it equals the base doping concentration, but increases very rapidly near the surface where it can have a value of several times 10^{20} cm^{-3}. The phosphorus solid solubility limit (about 7×10^{20} cm^{-3} at 900°C), which is the maximum concentration of phosphorus that can be dissolved in crystalline silicon, is easily reached. For values above this, the dopant precipitates out of the silicon solution. Near the solid solubility limit, the surface is not really silicon anymore, but rather, a compound or alloy of silicon and phosphorus. In such regions of very high surface concentration, the minority carrier lifetime is destroyed. It can be as small as 0.1 ns, with a corresponding minority carrier diffusion length of 0.1 µm. In this case, the top surface layer becomes a dead *layer*, and makes no contribution to the power output of the cell.

Base resistivities for terrestrial cells are chosen between 3 and 0.1 Ω cm. These values correspond to boron doping concentrations of 4.5 x 10^{15} and 2.5 x 10^{17} cm^{-3}, respectively. The more heavily doped material causes increased ionized impurity scattering of the minority carriers, and thus, shorter diffusion length. This lowers J_{sc} and increases J_o as expected from the theory of the ideal diode; but, *empirically, V_{oc} is often increased in direct contradiction to the theory*[4]. On the other hand, very lightly doped material improves J_{sc}, but the small V_{bi} limits V_{oc}. Thus, in terms of base doping, there is a trade-off between V_{oc} and J_{sc}. The actual base doping concentration for maximum efficiency will also depend on the quality of the material. For very low defect density single-crystal material, the higher base doping concentrations are preferred. High-efficiency cells made from float-zone material use base resistivities around 0.2-0.4 Ω cm and have diffusion lengths well over 500 μm. For cast polysilicon, resistivities between 1 and 3 Ω cm are preferred, with diffusion lengths in the 50-200 μm range. However, as shown in the above reference, gettering and passivation steps can be used to produce high efficiency cells (16.3%) even with high base doping and high J_o. This observation is significant for commercial solar cell fabrication because the source of the feedstock (Czochralski ingot scrap) is often already highly doped when purchased by a PV manufacturer. It demonstrates how important various material processing enhancements are to cell efficiency. Interestingly, high efficiency space cells tend to use the n$^+$p configuration with 10 Ω cm bases to afford enhanced radiation hardness.

In any commercial silicon cell process, phosphorus diffusion is achieved by first forming a *phosphosilicate glass* on the surface of the wafer. The phosphorus is supplied by either a gaseous or liquid source. Gaseous sources include phosphorus trichloride oxide $POCl_3$ and phosphine PH_3, with $POCl_3$ being the more common. $POCl_3$ is a liquid at room temperature, but is introduced into the diffusion environment as a vapor in nitrogen (fig. 4.2-1). Liquid sources include phosphoric acid H_3PO_4 in solution with organic solvents such as ethanol, isopropanol, and ethyl acetate. Gaseous sources are used in a tube furnace environment. Liquid sources are more versatile, as they can be sprayed or spun on the wafer at room temperature for subsequent diffusion in a belt furnace. Screen-printed phosphorus pastes have also been demonstrated for use in belt furnaces, but are infrequently used.

For gaseous diffusion with $POCl_3$, p-type silicon wafers are loaded into a quartz boat which is slowly moved into the middle of a fused quartz tube in a resistance-heated horizontal furnace. The boat, which can hold up to 50 wafers, is moved slowly into the tube so that the wafers will not suffer large temperature gradients and warpage. Tubes are periodically cleaned with HCl vapor in an N_2 stream. The vapor forms chlorides with metal impurities which are then flushed out of the tube. Furnace temperature for

Fig.4.2-1   Tray of wafers being loaded into a diffusion furnace. (*Photo courtesy Solarex*)

the diffusion is held at about 900°C, with a variation across the length of the boat of not more than 1°C. Close temperature control and tube cleanliness are important to minimize diffusion variations and contamination with metal impurities. During the loading and unloading procedure, N_2 gas flows through the tube to prevent oxidation and unwanted diffusion of impurities from the outside air. The diffusion step is the cleanest part of the fabrication process. Otherwise, solar cell fabrication is remarkably insensitive to contaminate-rich environments that would be disasterous in the fabrication of integrated circuits. Several minutes after being placed in the middle of the tube, the wafers reach the furnace temperature and the diffusion is started. Nitrogen is used as a carrier gas and bubbled through a flask containing $POCl_3$ liquid. The nitrogen becomes saturated with the liquid and is introduced into one end of the quartz tube. Oxygen is introduced separately into the tube. The gas mixture flows toward the far end of the tube where it can be vented and safely disposed. As the mixture of $POCl_3$-saturated N_2 and O_2 flows down the length of the tube and over the wafers, $POCl_3$ is oxidized in the reaction

$$4POCl_3 + 3O_2 \longrightarrow 2P_2O_5 + 6Cl_2$$

with the oxide P_2O_5 adhering to the surface of the wafers as a glass. The diffusion of phosphorus from the glass into the silicon bulk proceeds from the reduction reaction

$$2P_2O_5 + 5Si \longrightarrow 4P + 5SiO_2$$

As silicon dioxide forms, it consumes part of the silicon surface. The evolving surface of the silicon moves below the original surface level, with the thickness of silicon consumed equal to 46% of the total thickness of the oxide. However, oxide grows slowly, about 20 nm for 30 minutes at 900°C, and does not significantly interfere with the ability of P_2O_5 to reach the silicon surface if the entire diffusion is not too long.

Between spin-on and spray-on approaches for liquid phosphorus sources, the spray-on approach is more popular because it is amenable to use in a semicontinuous belt-furnace environment. Additionally, for non-circular wafers, use of a spin-on source results in a non-uniform glass layer. For a belt-furnace diffusion[5], the wafers are loaded onto a liquid-absorbent conveyor belt which goes into a dopant spraying chamber. The wafers are then passed onto another conveyor belt that goes through a diffusion furnace. The dopant spraying chamber can include a paper blotting mechanism so that any spray that reaches the edges or back side of the wafer are removed before the wafers go into the furnace. By doing this, there will be no diffusion of dopant on the wafer edges, thus obviating edge etching to prevent a short between the front and back sides. Belt furnace diffusion has four main steps: dopant solution spraying at room temperature, drying of the liquid to a uniform and thin layer, organic burnoff for several minutes at between 300° and 500°C to remove residual organic constituents from the solvents, and finally, heating to between 880° and 950°C in an oxygen ambient. The last step produces a phosphosilicate glass on the surface of the wafer, and the subsequent diffusion of phosphorus into the silicon bulk. A typical 30 minute diffusion into 1 Ω cm p-type material at 900°C produces a junction depth of about 0.4 μm.

When a phosphosilicate glass is deposited on a silicon substrate, phosphorus immediately starts diffusing into the bulk. The oxidation of $POCl_3$ is faster than the reduction of P_2O_5. Consequently, the glassy layer builds up and constitutes a layer with a uniform concentration of phosphorus. This is a predeposition. The resulting dopant profile is characterized by a shallow junction depth and a high surface concentration. As the process proceeds, SiO_2 starts to consume the silicon surface. If the dopant gas is turned off while the wafers are allowed to sit in the furnace with oxygen flowing, the predeposition will be terminated and the diffusion of dopant from the constant quantity already in the bulk will proceed as a drive-in. Under these circumstances, a layer of SiO_2 continues to grow. Eventually, it tends to seal the dopant into the bulk while it masks the transport of additional phosphorus from the glass to the silicon surface. Alternatively, to insure the termination of the predeposition, the wafer can be removed from the furnace and de-glazed by dipping in 10% HF solution for about a minute. Hydrogen fluoride preferentailly etches the oxide. The surface appears hydrophobic when the glass is completely removed. The wafer is then put back into the furnace for the drive-in. Drive-ins produce a deeper junction depth and a lower surface concentration. The junction depth is defined as the depth where the phosphorus and boron concentrations are equal. In general, for any impurity diffusion process, the junction depth increases and the surface concentration decreases with time.

Commercial terrestrial cells require high surface concentration for the industry standard screen-printed silver paste metallization by which the front contacts are made. In this process, a paste containing about 3% glass frit in a silver powder slurry is painted on the front surface through a stencil. While screen printing introduces three times as much series resistance as evaporated Ti/Pd/Ag metallizations used for high-efficiency cells, it has the great virtue of being fast, inexpensive, and amenable to automated processing. However, unlike evaporated metallizations, screen printing requires high surface concentration (sheet resistance of 40 to 50 Ω/$\square$) for acceptable ohmic contact, and deep junctions (at least 0.4 μm) to avoid shunting the junction. A short predeposition (875-925°C for 30 min) is sometimes sufficient. To achieve a deeper junction that will allow more insurance against junction penetration by the top metallization, a short drive-in (~ 15 min) at the same temperature is done in the same diffusion environment by simply turning off the dopant source. The relatively deep junction and dead layer near the surface degrade blue response.

Space cells require enhanced blue response to take advantage of the larger blue fraction outside the atmosphere. High efficiency terrestrial cells also benefit by improving blue response. This mandates both a shallow junction depth to allow penetration of blue light to the base as well as absence of a high recombination layer (*dead layer*) in the emitter where ultraviolet light is absorbed. A shallow junction can be achieved by a short predeposition, but yields a high surface concentration and dead layer. The dead layer can be removed by a uniform etchback with plasma etching or wet chemistry so that the junction depth is about 0.2 μm. Plasma etching of emitters with SF_6 has been investigated with limited results[6]. With wet chemical etching, it is difficult to obtain uniformity of shallow depth over the area of a wafer. Over-etching could result in a metal shunt through the junction, even with evaporated metal. If the emitter is to be uniformly and reliably etched back[7], a very slow isotropic etch solution is used, for example, $HNO_3:H_2O:HF = 100:100:1$. A 20 min etch in this solution reduces a deep 16 Ω/□ diffusion to a shallow layer with 80 Ω/□. An alternative approach to improving blue response is the formation of a *selective* emitter (fig. 4.2-2). This type of structure has a deep junction and low sheet resistance under the contact fingers, and a shallow junction and higher sheet resistance between the fingers. One approach uses a

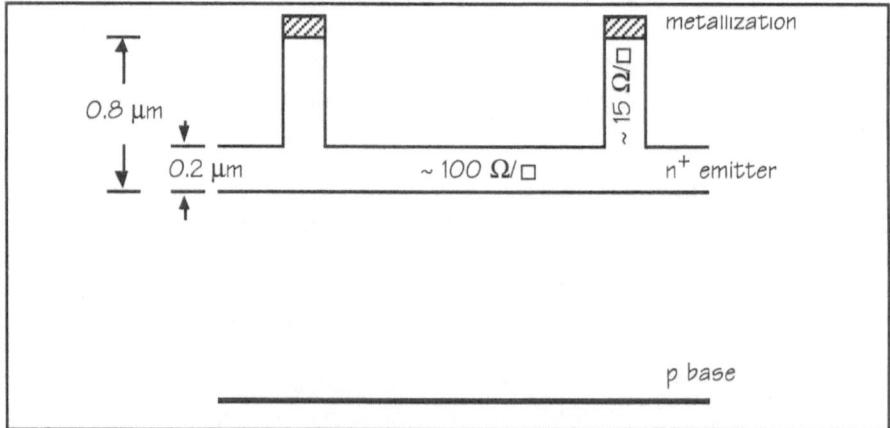

Fig. 4.2-2   Selective emitter structure for enhanced blue response.

predeposition at 875°C for 6 min and drive-in at 875°C for 70 min to yield a 0.8 μm junction. Next, Ti/Pd/Ag contact fingers are deposited by evaporation. The area between the fingers is then preferentially etched several tenths of a micron with CF_4 plasma with the metal fingers serving as a self-aligned mask for the silicon directly underneath. The result is an emitter structure which is about 0.2 to 0.3 μm thick in between the fingers and 0.8 μm thick beneath the fingers. Sheet resistance between the fingers is about 100 $\Omega/\square$, and about 15 $\Omega/\square$ under the fingers. The shallow junction in between the fingers allows enhanced absorption of blue light in the base, while the relatively low surface concentration decreases the emitter recombination current. Formation of a selective emitter has also been demonstrated with a process that achieves the phosphorus diffusion, backside Ag/Al metallization, and front-side metallization all by screen printing procedures[8]. An interesting advantage of selective emitter formation is that by removing the dead layer, it allows very effective use of hydrogen passivation. Diffusion of hydrogen into the bulk, either by plasma-enhanced chemical vapor deposition of silicon nitride or by thermal annealing in a hydrogen ambient (forming gas) tends to passify bulk and surface states. Passivation with PECVD SiN_x is common because the nitride also serves as an anti-reflection coating. However, the plasma treatment damages the surface and creates another dead layer. In selective emitter processes, this is avoided by performing the nitride deposition before etching back the emitter.

Diffusions occur more or less uniformly across the surface of the wafer or ribbon and the transport of impurity into the bulk can be treated as one-dimensional. The diffusion equation is obtained by considering the impurity flux (atoms per unit area per unit time) across a thin rectangular box straddling the surface of the wafer (fig. 4.2-3). The faces parallel to the wafer surface have unit areas. For a predeposition, the top surface of the box can be thought of as being the phosphosilicate glass layer. The thickness of the box is Δx, where x is measured downward from the top face of the box. The flux $F(x)$ of impurity atoms and the average concentration C_{AV} are related by

$$\frac{\partial C_{AV}}{\partial t} \Delta x = F(x) - F(x + \Delta x) \tag{4.2-1}$$

As $\Delta x \rightarrow 0$, $C_{AV} \rightarrow C(x,t)$ and

$$\lim_{\Delta x \rightarrow 0} \frac{F(x) - F(x + \Delta x)}{\Delta x} = -\frac{\partial F(x)}{\partial x} \tag{4.2-2}$$

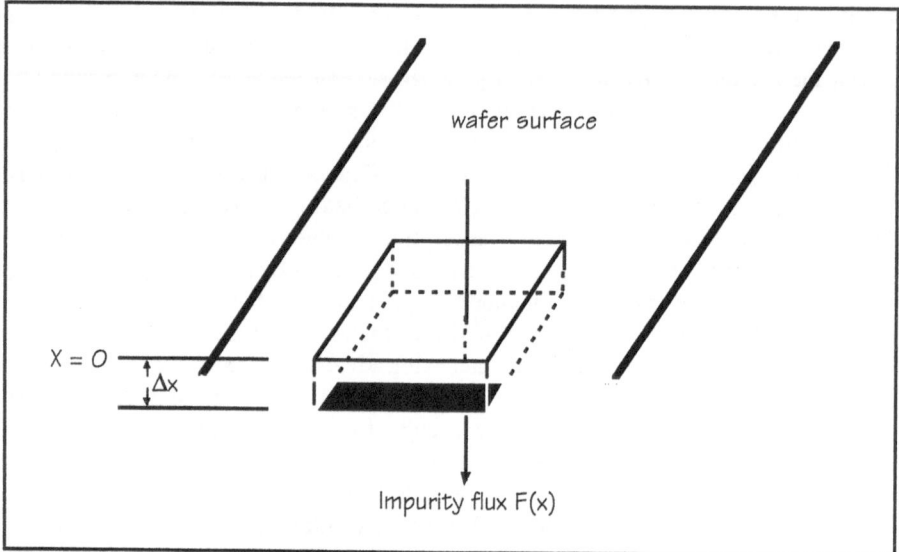

Fig. 4.2-3   Unit area box straddling diffusion surface.

The minus sign indicates that the impurity atoms are flowing out of the surface and into the bulk. Thus

$$\frac{\partial C(x, t)}{\partial t} = - \frac{\partial F}{\partial x} \qquad (4.2\text{-}3)$$

Intuitively, the flux of atoms from the phosphosilicate glass into the bulk is expected to increase with the negative gradient of the concentration. The flux is related to the gradient of the concentration by the diffusivity of the *impurity*, D. Impurity diffusivity is dependent on temperature, and in the case of phosphorus, also dependent on surface concentration. Boron diffusivity is independent of surface concentration. For the one-dimensional case

$$D \frac{\partial C(x, t)}{\partial x} = - F(x) \qquad (4.2\text{-}4)$$

The minus sign indicates that the impurity atoms move in the direction of decreasing concentration. Substituting eq. (4.2-4) into (4.2-3) yields the one-dimensional diffusion equation:

$$\frac{\partial C(x, t)}{\partial t} = D \frac{\partial^2 C(x, t)}{\partial x^2} \tag{4.2-5}$$

With constant surface concentration C_s, the boundary and initial conditions for the predeposition are

$$C(0,t) = C_S$$

$$C(\infty,t) = 0$$

$$C(x,0) = 0$$

The solution[9] is given by the complementary error function

$$C(x, t) = C_S \, erfc \left(\frac{x}{2\sqrt{Dt}} \right) \tag{4.2-6}$$

where

$$erfc(x) = 1 - erf(x) = 1 - \frac{2}{\sqrt{\pi}} \int_0^x e^{-y^2} \, dy$$

The quantity $2(Dt)^{1/2}$ represents a characteristic diffusion length for the impurity. Figure 4.2-4 shows curves of normalized concentration versus distance for several phosphorus diffusion times. The curve for $t = 15$ min indicates the large emitter gradient resulting from a short diffusion. The area under each curve in fig. 4.2-4 is given by

$$Q_A(t) = \int_0^\infty C(x, t) \, dx$$

$$= \frac{2}{\sqrt{\pi}} \sqrt{Dt} \, C_S \tag{4.2-7}$$

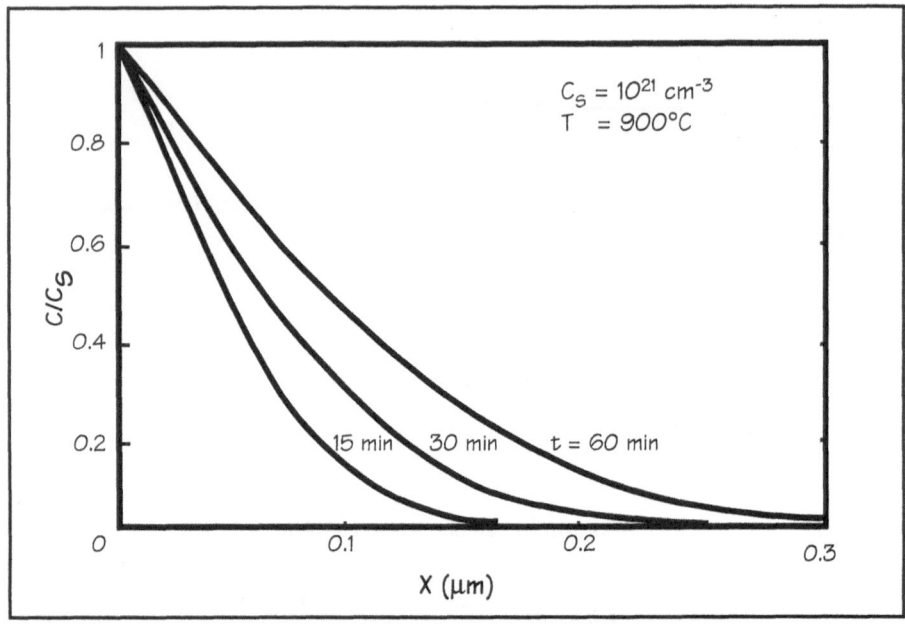

Fig. 4.2-4   Complementary error function (erfc) showing normalized
concentration versus distance into the wafer for several phosphorous diffusion times.

and represents the number of impurity atoms per unit area after time t. For any given
diffusion temperature, the total quantity of impurity diffused in time t is dependent on
the surface concentration which is effectively *limited to the solid solubility of the impurity.* This is significant because the solid solubility is only weakly dependent on temperature, and thus, the total quantity of impurity diffused into the semiconductor will
be relatively insensitive to the partial pressure of the dopant gas. The necessary limitation of the surface concentration to the solid solubility allows good control of the diffusion and predictable results from one batch of cells to the next.

Since the emitter is diffused into a uniformly doped substrate, the junction depth x_j is
given by

$$C_{BC} = C_o \, erfc \left(\frac{x_j}{2 \, \sqrt{(Dt)_{predep}}} \right)$$

(4.2-8)

where C_{BC} is the background concentration of the substrate, and C_o is the solid solubility limit of the dopant at the diffusion temperature. If the wafers are removed immediately after the predeposition, the rapid cool-down locks the surface concentration at C_o.

The predeposition deposits a quantity Q of impurity in the bulk. It can be followed by a short drive-in to bring the surface concentration below the solid solubility and increase the junction depth. During a drive-in, the impurity quantity Q is assumed *trapped* in the bulk. Thus, the gradient of impurity across the surface is zero. If the drive-in occurs in oxygen so that a masking oxide grows during the process, this can be a very good assumption. Additionally, the thermal oxide can serve as a surface passivant. For a drive-in, the boundary conditions on C(x,t) are

$$\nabla C(0,\ t) = 0$$

$$C(\infty,\ t)\ =\ 0$$

The initial condition for the drive-in (t = 0) is the final condition of the predeposition:

$$C(x,\ 0)\ =\ C_o\ erfc\left(\frac{x_j}{2\ \sqrt{(Dt)_{predep}}}\right)$$

The drive-in diffusion solution is readily obtained with the use of a delta-function approximation to the initial impurity profile, and is given by

$$C(x,\ t)\ =\ \frac{Q}{A\ \sqrt{\pi\ Dt}}\ e^{-x^2/4Dt} \tag{4.2-9}$$

where t is now the drive-in time. The delta function approximation is very good for a long drive-in where the junction depth is much greater than that rendered by the predeposition. This is often not the case for solar cells. Also, if the phosphosilicate glass is not removed before the drive-in and excess phosphorus has precipitated on the surface, the expression in eq. (4.2-9) becomes even less valid. Regardless of the ability to produce a simple and accurate analytic expression for the drive-in profile, the drive-in process is very controllable.

Fig.4.2-5   Cells exiting belt furnace. (*Photo courtesy Solarex*)

As indicated above, commercial terrestrial cells would benefit from an inexpensive process that allows both shallow junctions (for improved blue response in the base) and high surface concentrations (for low interfacial contact resistance with screen printing). In fact, this requirement is of great importance to the silicon integrated circuit industry where short-channel CMOS circuits require heavily doped sources and drains with shallow junctions. To accomplish this, the silicon IC industry uses ion implantation. This is a method that accelerates charged ions of the dopant to high energies (150-250 keV) and showers these particles upon the silicon surface. The kinetic energy of the particle and the angle of incidence controls the depth of the implant. Arsenic is the dopant of choice for n-type implants. Arsine gas, AsH_3, is the dopant source. Ion implantation produces surface damage due to the displacement of silicon atoms as the arsenic ions collide with the lattice, but this is remedied by the simultaneous annealing of the lattice during the procedure. A considerable amount of research has been done on the application of ion implantation for creating solar cell emitters. Unfortunately, the low-throughput rate and large capital equipment expense of ion implanation make this approach unsuitable for solar cells. The same can be said for phosphorus doping through neutron transmutation of silicon into phosphorus at the top surface of the wafer. It appears that diffusion from a gaseous or liquid source will remain the method of choice in the crystalline silicon PV industry for many years to come.

After diffusion, the top phosphosilicate glass layer is removed by dipping the wafer in buffered 10% HF solution for about a minute. Hydrogen fluoride solution etches SiO_2 (and the accompanying P_2O_5) but only negligibly etches the silicon substrate. Diffusion in a tube produces phosphorus doping on both sides of the wafer and on the edges. This can also happen in a belt furnace if the liquid source is not removed from the back surface and edges beforehand. A phosphorus diffusion on the back surface is not problematic because the phosphorus-doped silicon becomes part of the back-side aluminum metallization. However, phosphorus diffusion on the edges must be removed to prevent shorting the emitter to the back side. This is accomplished by stacking the wafers in a coin configuration and then wet etching in a buffered HNO_3 solution to etch off the silicon on the edges. Alternatively, the stacked wafers can be etched by a CF_4 plasma.

The two-sided diffusion produced by a tube furnace enhances the gettering of metal impurities from the bulk of the wafer. During the routine phosphorus diffusion, metal impuities diffuse toward the glass on both sides of the wafer. Double-sided tube diffusion results in slightly improved performance at a cost of more time and lower throughput from loading and unloading wafers. Table 3 shows a comparison of the advantages and disadvantages of the tube furnace and belt furnace approaches[10].

COMPARISON OF TUBE FURNACE AND BELT FURNACE DIFFUSION APPROACHES

	Control of C_o and x_j	P Gettering of Impurities	Control of Contaminants	Equipment Cost	Wafers per hour	Energy (kWh/wafer)
Tube	Better	Better	Better	Worse	200	5 - 10
Belt	Worse	Worse	Worse	Better	1000	5 - 10

Table 3   Comparison of tube furnace and belt furnace diffusion approaches

4.3 PERFORMANCE ENHANCEMENTS

Aside from the generic limitations discussed in section 3.5, there are specific parametric limitations that place an upper limit on silicon cell efficiency. For one-sun illumination of silicon cells that include the ideal structural qualities of complete light trapping and highly passivated bulk and surface regions so that every incoming photon above the bandgap contributes to the output power, the calculated values for V_{OC}, J_{SC}, and FF are about 850 mV, 44.5 mA/cm^2, and 0.89, respectively[11]. This yields an efficiency limit of about 33%. Other calculations incorporating light trapping predict about 30% maximum efficiency[12]. Calculated predictions of efficiency are often based on the assumption that design and material characteristics controlling these parameters can be independently optimized. Clearly, this is not the case, as the optimization of one parameter interferes with the optimization of the others. An example is the relationship between J_{SC}, V_{OC}, and cell thickness. As the thickness increases, J_{SC} increases as more and more photons are absorbed and produce pairs. Eventually J_{SC} saturates as the median distance of a photogenerated minority carrier significantly exceeds a diffusion length from the junction. The saturation is expected for thicknesses between 100 and 300 microns, and the prediction is experimentally verified. Meanwhile, conventional theory predicts V_{OC} will decrease with thickness because of increased bulk recombination[13]. Conversely, V_{OC} is expected to increase significantly for thin cells in those cases where surface recombination is less than bulk recombination. Very high values of V_{OC} would be expected for silicon cells less than 100-μm thick. This expectation has been verified in a 47-μm device yielding 699 mV[14]. However, in other cases, experimental evidence for this prediction is not observed[15]. The discrepancy with theory is attributed to insufficiently passivated back surfaces in very thin cells, so that surface recombination becomes large compared to bulk recombination. Surface recombination dominance becomes particularly severe when the bulk lifetime is large to begin with, for example, as seen in float-zone material. Interestingly, the silicon cell with the highest open-circuit voltage as of 1996 (709 mV) was 400-μm thick[16]. This leaves open the possibility that V_{OC} does not inherently increase as the thickness decreases and that mechanisms other than surface recombination may play a role in the frequent observation of less-than-predicted V_{OC} in thin cells. In any case, regardless of the possibility of unconventional device physics, the example illustrates the *close interaction between processing and performance*. Commercial cell manufacturers seek enhancements that manipulate V_{OC}, J_{SC}, and FF so that power per unit cost is increased.

The baseline commercial crystalline silicon cell consists of only a pn-junction with screen printed metallization and anti-reflective coating on the front side and aluminum contact metallization on the back side. The front and back metallizations are fired to allow some dissolution of the metal into the silicon and produce a good ohmic contact. Baseline cell AM1.5 efficiency is about 10% for polysilicon and 11.5% for CZ single crystal. With added cost, this is improved considerably by several enhancements (Table 4). These include texturing of the front side for light trapping, double layer anti-reflec-

PERFORMANCE ENHANCEMENTS DURING CELL FABRICATION

Enhancement	Purpose
Pyramidal texturing of front side	Light trapping
Surface passivation with thermal SiO_2	Lower surface recombination velocity
Forming gas anneal	Lower surface & grain boundary recombination
Passivation with PECVD oxides & nitrides	Lower surface and bulk recombination rates
Formation of back surface field	Increase effective diffusion length
Phosphorus gettering from front & back sides	Remove metal impurities from bulk
Aluminum gettering from back side	Remove metal impurities from bulk
Etch-back of emitter between contact fingers	Increase blue light penetration to base
Anti-reflection coating(s)	Increase transmission coefficient
Evaporated metallization	Lower C_s, lower $R_{contact}$, lower R_s

Table 4   Performance enhancements during cell fabrication

tive coating to produce a broad spectrum quarter wavelength matching section on the front side, hydrogen diffusion to lower surface and bulk recombination rates, formation of a back-surface field to turn around minority carriers in the base that approach the rear contact, extrinsic gettering (i.e., attraction of impurities from the bulk to the front or back surface) to increase minority carrier diffusion length, evaporated metallizations to obviate the highly-doped emitter surface concentrations associated with screen printing, and rarely, selective emitters where the area between the contact fingers is etched back several tenths of a micron to achieve improved blue response in both the emitter and base. It is the accumulation of relatively minor process enhancements, all in some way associated with the integrated circuits industry, that allows the crystalline silicon technology to maintain its predominance in photovoltaics.

A. Light Trapping

Silicon's high index of refraction[17] varies from n = 3.5 at 1.1 μm (infrared) to n = 5.6 at 0.4 μm (violet), and produces a large reflection coefficient across the solar spectrum[18]: 34% at 1.1 μm increasing to 54% at 0.4 μm. Any mechanism that increases the coupling of light into the silicon surface improves the short-circuit current. There are two approaches: texturing of the surface to produce multiple reflections back into the silicon and creation of a matching section to lower the reflection coefficient (fig. 4.3-1). Chemical texturing is performed before junction formation; anti-reflection coatings are applied after junction formation.

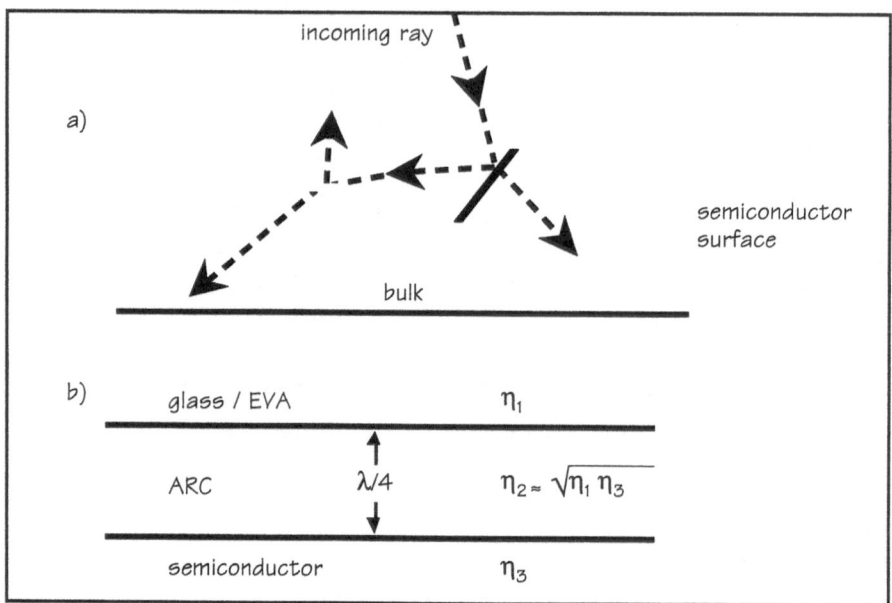

Fig. 4.3-1   Light trapping approaches: (a) chemical texturing and (b) anti-reflection coating.

Atomic packing density varies with planar orientation. In the diamond structure, the (110) plane is less dense than the (100) plane. This, in turn, is less dense than the (111) plane. For epitaxial deposition growth and for surface etching, there are two competing rates. One is the rate of the kinetics of the reaction at the surface. The other is the rate at which reactants diffuse to the surface. The slower of the two rates controls the overall rate of growth or etching. Orientation dependent etchants are distinguished by the property that the rate of reaction at the surface controls the overall etch rate. For these etchants, the (100) plane is expected to etch at a faster rate than the (111) plane provided there is no intermediate oxide formed to retard the arrival of etchant molecules at the surface. An aqueous solution of propanol and KOH[19] provides an example. At 80°C, (100) planes etch about two orders of magnitude faster than (111) planes. For a (100) silicon wafer, KOH/propanol/water solution creates four-sided pyramids consisting of sections of (111) planes which form internal angles of 54.7° with the (100) surface. A solution consisting of $KOH:CH_3CH_2CH_2OH:H_2O = 24:13:63$ (w/w) etches (100) planes at 0.6 μm/min and (111) planes at 0.006 μm/min. The overall reaction is

$$Si + H_2O + 2KOH \rightarrow K_2SiO_3 + 2H_2$$

where the silicate is water soluble. Similarly, the <110> direction is anisotropically etched by an aqueous solution of KOH (35% by weight) at 80°C. Etching is 600 times faster in the <110> direction than in the <111> dirction. For certain etchants, the degree of isotropy is sensitive to the concentration of the components. An example is the water/isopropanol/NaOH system. While a 300g NaOH/l solution at 80°C etches silicon isotropically to render a polished surface, a 20 g/l solution at 90°C etches anisotropically to expose (111) planes. Chemical texturing has little usefulness for polycrystalline material where the grains are randomly oriented. It is reserved for (100) or (110) single-crystal material.

Aside from decreasing the net reflectance from the surface, texturing has the advantage of increasing the pathlength of the light through the silicon (due to the low angle of incidence after refraction through a pyramid). Minority carriers are then generated slightly closer to the collecting junction than if the photon entered the surface orthogonally. This tends to *decouple* the performance of the cell from the minority carrier diffusion length in the base, and improves J_{sc}. If the back side is specular, the low angle of incidence will allow red light to undergo multiple passes and improve the red response of the cell. On the negative side, chemical texturing requires wet processing, which is time consuming. It is restricted to more expensive high performance cells.

Anisotropic etchants are contrasted with certain solutions of HF and HNO_3 (with CH_3COOH or water as a diluent) where silicon is isotropically[20] etched due to the dependency of the process on the formation of an SiO_2 intermediate. The overall reaction is the result of two competing processes[21], one where silicon is oxidized by nitric acid and the other where silicon dioxide is dissolved by hydrofluoric acid:

$$18HF + 4HNO_3 + 3Si \rightarrow 3H_2SiF_6 + 4NO + 8H_2O.$$

Nitric acid etches silicon and produces silicon dioxide. The accumulating oxide masks the silicon substrate and prevents further etching until it is dissolved by hydrofluoric acid. When HF is in limited supply, dissolution of the oxide is the rate-controlling process. Delay caused by the masking action of the oxide allows the etching of the (111) plane to catch up with that of the (100) plane and the etching is isotropic. The HF:HNO$_3$:diluent ratio controls the degree of isotropy and rate of substrate etching. There is a large range of HF to HNO$_3$ ratios which remove small surface damage features and produce a polished highly planar surface[22]. A ratio of HF:HNO$_3$:CH$_3$COOH = 8:75:17 (v/v) etches at 5 µm/min at 25°C to yield a polished surface regardless of the orientation. A lower ratio of HF to HNO$_3$ slows the overall reaction considerably. For 1:40:15, the etch rate is only 0.15-0.20 µm/min. One of the by-products of the overall reaction is water. As the reaction continues, the solution becomes increasingly water diluted. Nitric acid is more ionized in water than in acetic acid. Thus, acetic acid is preferred to water as a diluent because it lends enhanced stability to the reaction. Etch rate for the HF/HNO$_3$/CH$_3$COOH system is also dependent on the impurity concentration. A 1:3:10 solution etches p$^+$ and n$^+$, but stops at p$^-$ or n$^-$. This property allows lightly doped layers to serve as an etch stop in the processing of some structures. Other etch systems are both orientation and impurity dependent. Hot aqueous solutions of ethylenediamine with pyrocatechol preferentially etch silicon in the <100> direction and stop etching at a p$^+$ layer.

As light approaches a solar cell in a module, it passes from air into a four-layer sandwich: glass/EVA/anti-reflective coating/silicon cell. In general, for any sandwich structure, reflections occur at interfaces where there is an impedance mismatch between the materials, or alternatively, where there is a difference in the indices of refraction. The index of refraction for a given material is the speed of light in free space normalized to that in the material. The speed of light is given by c = 1/(µ$\in$)$^{1/2}$, where µ is the magnetic permeability and $\in$ is the electric permittivity. In the glass/EVA/ARC/silicon sandwich, µ ≈ 1, and the index of refraction reduces to n = ($\in$/$\in_o$)$^{1/2}$ for each layer, where $\in_o$ is the electric permittivity of free space. For sufficiently thin layers with very flat interfaces, multiple reflections establish an interference pattern which controls the reflection and transmission coefficients for the overall sandwich structure. Under these conditions, the optimization problem for complete transmission into the bottom layer (silicon) is amenable to an analytical solution. However, in a glass/EVA/ARC/silicon sandwich, the EVA (n = 1.48) is typically 0.46 mm thick, while the glass (n = 1.5) is about 3 mm thick. Additionally, large sheets of glass tend to have a wavy surface. As a result, the glass/EVA/ARC/silicon problem is approached as a three-layer problem with the glass/EVA treated as one layer with n = 1.5. From electromagnetic theory[23], for the three-layer (glass and EVA/ARC/Si) sandwich, the transmission of light into the silicon at a given wavelength λ is maximized when two conditions are satisfied:

(i) the index of refraction of the ARC is equal to $(n_{glass/EVA} \, n_{Si})^{1/2}$, and

(ii) the thickness of the ARC is equal to $ml/4$, where m is an odd integer.

As an example, a perfect impedance match is achieved at $l = 0.55$ µm with an ARC having a thickness of 0.1375 µm and an index of refraction of 2.27. The coating approximates a quarter-wavelength matching section between the EVA encapsulant and the silicon. The spectral response of a cell is maximized at that wavelength where the skin depth equals the junction depth. Thus, the optimal anti-reflection coating varies with base resistivity.

Common anti-reflection coatings for silicon are TiO_2 (n = 2.25) deposited by atmospheric-pressure chemical vapor deposition from an organometallic precursor, and SiO_2 (n = 1.45) and Si_3N_4 (n = 2.0 to 2.2) deposited by plasma-enhanced chemical vapor deposition. Since neither PECVD nitrides and oxides are necessarily stoichiometric, they are sometimes represented as SiN_x and SiO_x. The film density and index of refraction varies with the mole fraction x, which depends on the deposition parameters. Precursors for silicon nitride are ammonia (NH_3) and silane (SiH_4); precursors for silicon dioxide are silane and oxygen. Films are also deposited by sputter deposition, a physical vapor deposition process in which a target surface is bombarded by energetic non-reactive ions, usually argon. Target atoms are kinetically ejected and condense on the substrate to form a uniform film. Anti-reflective coatings reduce the average reflection in the range of wavelengths 0.4 µm to 1.1 µm to between 6 and 10% absolute. This does not include the reflection from the air/glass interface. A coating on the glass with n = 1.23 matches glass to air and reduces the average overall reflection down to about 4% absolute. The application of two different AR films on the silicon surface improves the optical match by spreading the already improved transmission curve over a larger section of the solar spectrum. For example, 60 nm TiO_2 and 110 nm SiO_2, reduces the average reflection (not including the air/glass interface) to 3% absolute over the silicon response range[24]. TiO_2/Al_2O_3 is highly effective and popular for concentrator cells. Other combinations are 59 nm SiN_x (n = 2.3)/95 nm SiO_2 , and 55 nm ZnS/110 nm MgF_2 (n = 1.4)[25]. Like many other processes, AR films are applied in a continuous moving belt environment (fig. 4.3-2).

B. Passivation

Passivation refers to mechanisms that lower recombination rates either in the bulk, along grain boundaries, or at the surface. Performance is degraded by any current of minority carriers flowing into, and recombining at, a point defect or interface. The

Fig.4.3-2   Cells emerging from an anti-reflective coating furnace. (*Photo courtesy Solarex*)

resulting loss of excess minority carriers contributes to the total saturation current of the cell I_O, and decreases V_{OC} and I_{SC}. Major contributions come from recombination at the back surface contact, front contacts, emitter surface, bulk defects in the base, and grain boundaries (for polycrystalline material). At these locations, disruption of the crystal periodicity introduces intermediate energy states in the bandgap that facilitate recombination. The resulting concentration gradient of minority carriers drives the re-combination current. At surfaces and grain boundaries, passivation presumably ties up dangling (uncoordinated) covalent silicon bonds, thereby decreasing the density of states in the bandgap. Passivation takes the form of thin film deposition of oxides on the uncontacted areas of the emitter and rear surfaces, and introduction of hydrogen into bulk and grain boundary regions through deposition of silicon nitride or through form-ing gas (10% H_2, 90% N_2) annealing. Because passivation is performed after the diffu-sion, high temperature depositions must be avoided to prevent alteration of the diffu-sion profile. This can be accomplished by PECVD, a method characterized by low temperatures (250-450°C)[26] followed by short annealing. Internal quantum efficiency provides a convenient method for differentiating the region of passivation. Improved blue response indicates diminished surface recombination velocity at the emitter surface. Improved red response indicates passivation of point defects and grain boundaries deep in the base.

Thermally grown silicon oxides are effective passivants and anti-reflection coatings; TiO_2 is commonly used because it is easily deposited by a pyrolytic spray or APCVD procedure. Non-stoichiometric films of silicon nitride, sometimes in combination with a PECVD oxide to simultaneously serve as a dual-layer anti-reflection coating, are increasingly popular because of their effectiveness[27]. Reaction gases are 2% to 3% silane in N_2 and pure ammonia. An effective deposition is 10 nm thermal oxide underneath 60 nm SiN_x underneath 95 nm PECVD SiO_2[28],[29]. This is followed by an anneal in N_2. The refractive indices of the SiN_x and SiO_2 PECVD films are about 2.3 and 1.45, respectively. A PECVD SiN_x film and post deposition anneal strongly passivates both the emitter surface and bulk defects in those materials (e.g., EFG ribbon) that are susceptible to the nitride-induced hydrogenation[30]. A PECVD SiN film and post-deposition anneal passivates both the emitter surface and defects in the base. The annealing conditions are critical to the effectiveness of the passivation even though the effectiveness varies considerably with the type of substrate. An anneal at 350°C for 20 minutes raises the efficiency of a simple baseline cell made from cast polycrystalline material about 1% absolute. Annealing at higher temperatures and for shorter periods of time (accomplished by rapid thermal annealing with infrared lamps) further decreases the component of saturation current contributed by the emitter so that a 650°C anneal for 1.5 minutes produces approximately 2% absolute efficiency improvement. However, the higher temperature degrades the internal quantum efficiency below 500 nm. This is attributed to absorption of light below 500 nm by the nitride film due to annealing-induced depletion of the film's hydrogen composition. Another method that effectively passaivates bulk defects is forming gas annealing following backside aluminum alloying. The cell is heated above the eutectic temperature (577°C) to form a silicon/aluminum solution, followed by a short heat treatment at 400 - 600°C in the forming gas. Efficiencies are raised about 3% absolute compared to cells that do not receive the annealing[31].

For both PECVD SiN_x deposition and forming gas annealing after aluminum alloying, the mechanism of passivation is attributed to hydrogen transport into the bulk[32],[33] facilitated by the injection of surface damage induced vacancies. Vacancies increase the solubility of hydrogen at the surface and increase the rate at which molecular hydrogen dissociates into hydrogen ions. In the presence of molecular hydrogen, vacancy injection from the surface damage caused by either PECVD nitridation or aluminum alloying will allow hydrogen ions to penetrate deep into the bulk and tie up uncoordinated bonds. For PECVD nitridation, hydrogen is incorporated in the film from the SiH_4 and NH_3 precursors. The overall reaction for the stoichiometric deposition[34] of nitride at high temperatures is

$$3SiH_4 + 4NH_3 \rightarrow Si_3N_4 + 12H_2$$

But at the low temperatures characteristic of PECVD, the deposition is non-stoichiometric and the nitride is represented as Si_xN_yH, where $x \approx y$ and the hydrogen content is about 20 atomic percent. Similarly, for forming gas annealing, the hydrogen comes from the ambient gas. Consistent with the vacancy assisted hydrogen diffusion model for passivation is the observation that a 600°C post-forming gas treatment in pure N_2 defeats the forming gas effect. This is attributed to the out-diffusion of hydrogen in the absence of the hydrogen-rich surface provided by the forming gas ambient. On the other hand, a 600°C post nitridation anneal in pure N_2 does not have a deleterious effect on cell efficiency, presumably because the hydrogen-rich surface is held in place by the relatively impermeable SiN_x layer. An additional observation consistent with the vacancy assisted model is that those silicon growth technologies that are characterized by high concentrations of vacancies will have relatively high hydrogen diffusivities. Such materials include laser-recrystallized silicon ribbons where the high growth rate (and subsequent fast cooling rate) quenches large concentrations of vacancies, and also edge-defined film-fed growth and edge supported ribbons where the high growth rate and high carbon concentration from the die both favor vacancies. Materials that are grown slowly, e.g., float-zone, Czochralski, and to a lesser extent, cast silicon, have low concentrations of vacancies and low hydrogen diffusivities. Czochralski material, in particular, has a low hydrogen diffusivity because the high concentration of oxygen from the silica crucible reduces the vacancy formation rate in the ingot. Consequently, ribbons are expected to benefit more from post-Al alloying forming gas annealing or SiN_x passivation than does Czochralski material. The performance of *experimental* edge supported ribbon, for example, has been greatly improved by a 70 nm SiN_x layer, with absolute efficiency increasing from about 11.7% (with TiO_2 ARC only) to over 14% (with SiN_x film).

C. Back Surface Field

Back surface field (BSF)[35] refers to a built-in electric field on the back side that deflects minority carriers and reduces the recombination rate at the back surface[36]. The back surface recombination velocity S_n can be reduced from an initial value of 10^7 cm/s to a passivated value as low as 300 cm/s. With an effective BSF, fewer minority carriers will be able to approach the back contact and recombine. The effectiveness is a function of the strength of the field and the width-to-diffusion length ratio. When $W/L < 1$, as, for example, in a 300-μm thick cell made from float-zone material with 1500-μm bulk diffusion length, a decrease in the back surface recombination velocity by several orders of magnitude will slightly improve minority carrier collection and significantly lower J_o. Effective back-surface fields increase V_{oc} by 20 mV or more and increase cell efficency by about 1.5% absolute. For $W/L \approx 2$, as in cast silicon, the improvement is less impressive, typically 0.5% absolute. For reasons of both cost (material economy) and performance (enhanced carrier collection), the trend in cell fabrication is toward thinner cell designs. Many crystalline cells are now about 200-250 μm thick. This increases the need for an effective BSF so that minority carriers can take advantage of the lower W/L ratio.

Back-surface fields take the form of a p^+p junction (high-low junction) formed either by boron diffusion from a borosilicate glass[37] or aluminum alloying from a screen-printed paste or evaporated layer on the back surface. Aluminum is a p-type dopant in silicon. The band diagram for the high-low junction (fig. 4.3-3) shows that the electric field at the p^+p junction points toward the p^+ layer. Minority electrons in the base are turned around by the field and have an improved chance of reaching the emitter/base

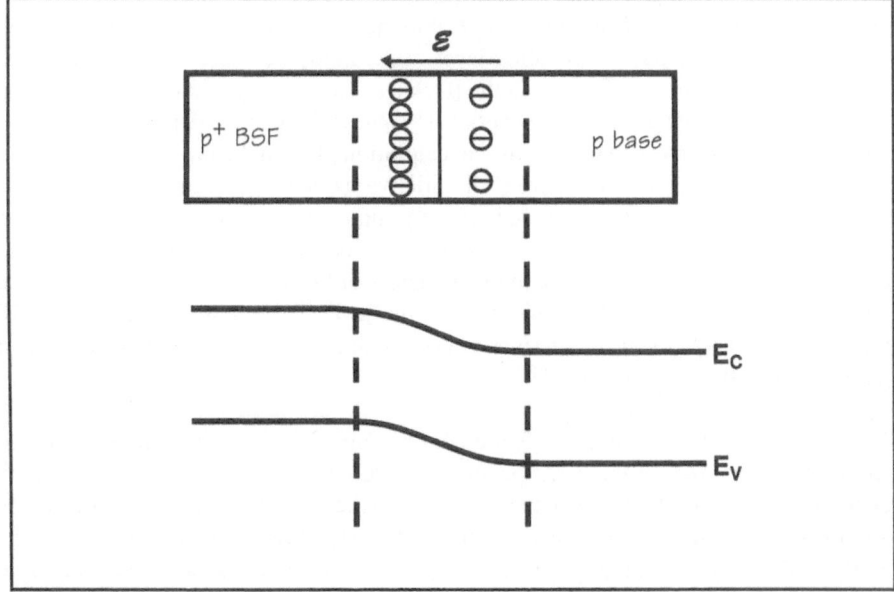

Fig. 4.3-3   Band diagram for high-low junction on back surface.

pn-junction. Thus J_{sc} is slightly improved. Most of the effect of the BSF, though, is seen in the lower value for J_0 and the significantly improved value for V_{OC}. Formation of a p⁺p junction through boron diffusion in a tube or belt furnace is problematic because the micron or more depth required for an effective diffused-boron BSF implies a long diffusion time which alters the top-side pn junction. Front-side phosphorus and back-side boron have been diffused simultaneously. Co-diffusion allows control of the phosphorus diffusion, but the different diffusivities of phosphorus and boron make the optimal time and temperature for the diffusions unequal. At the preferred temperature of about 900°C for the phosphorus diffusion, the phosphorus diffusivity is about five times that of boron[38]. Low backside surface recombination velocities have been achieved with co-diffusion, but the higher temperature requirement degrades the bulk lifetime.

Formation of a BSF by aluminum alloying is more common and yields higher performance than boron diffusion. Unlike boron diffusion which produces a shallow heavily-doped p⁺ layer, aluminum alloying produces a relatively low-doped very thick p⁺ layer. Several microns of the metal are applied by evaporation or paste to the back side, followed by rapid heating to 770-900°C and then cooling. Treatments of 850°C for 30 sec have produced back-surface fields with aluminum reaching a peak concentration at a depth of 10 μm. Short treatments are preferred to avoid diffusion of unintended impurities from the aluminum layer into the base of the cell. Not all of the metal necessarily becomes part of the alloy layer – there may be excess aluminum on the back surface after the alloying. Both the thickness and concentration of the Al-doped layer are critical to the BSF effectiveness[39]. Thickness of the layer depends on the total amount of Al in the paste, whereas, the concentration of Al in the layer depends mostly on temperature. Reduction of S_n from an initial value of 10^7 cm/s to a final value less than 10^3 cm/s has been demonstrated with a paste containing 4.7 mg Al/cm² and an alloying temperature of 770°C[24]. Reduction in the concentration of the paste produces an approximately linear reduction in the thickness of the alloy layer. Reduction of the alloy temperature reduces both thickness of the layer and doping density. The BSF process can be accompanied by a short oxygen soak to produce a simultaneous silicon oxide passivation on the front side[40].

In the alloying process, a liquid mixture whose composition is temperature dependent forms above the Al-Si eutectic temperature (577°C). From the Al-Si phase diagram[41], as the mixture cools down, the concentration of silicon in the liquid phase decreases and the excess silicon recrystallizes on the wafer. The layer contains aluminum at its solid solubility limit for that particular temperature. When the system cools down to the eutectic temperature, the remaining liquid turns to solid with the eutectic composition given by the phase diagram (about 12 atomic percent Si). The result is an Al-doped layer several microns thick and a subsequent back surface field. The back-surface field is formed by a regrowth process as the silicon solidifies, and not by diffusion of aluminum into the silicon[42]. A shortcoming is the low solid solubility of Al in Si. This limits aluminum concentration to 1-3 x 10^{18} cm^{-3} unless prohibitivily large processing temperatures or alloy times are used[43]. The resultant strength of the BSF can be too small to significantly lower S_n at the back side. Since the strength of the BSF is an increasing function of both the doping concentration and the thickness of the alloyed layer, S_n can still be brought to 10^3 cm/s by using a sufficiently thick alloy layer. This means using a high-Al concentration paste and a short high-temperature treatment. Problems include expense due to the thick Al paste and possible warpage in thin wafers due to thermal mismatch between the bulk Si and the alloyed layer. Ideally, a BSF producing a low value of S_n is formed by a thin highly doped p$^+$ layer. Recent work[35] has demonstrated S_n values of about 400 cm/s with back surface fields formed by 1%-boron-doped aluminum pastes on 1 Ω cm Czochralski Si material. Peak alloy temperatures were 850°C for less than 1 minute, a characteristic which makes the process ideal for a rapid thermal annealing environment. This approach produces peak dopant concentrations of 3 x 10^{19} cm^{-3} at a depth of 3 µm into the alloyed layer, as opposed to an Al-only approach that produces a peak dopant concentration of 3 x 10^{18} cm^{-3} at a depth of more than 6 µm into the alloyed layer (short duration peak temperatures of 85°C in both cases). The Al/B-doped BSF lowers S_n to 400 cm/s compared to 500 cm/s for the Al-only BSF, even though the peak dopant concentration occurs at less than half the depth. Experimental processes that use boron and evaporated aluminum have also been reported[44]. Aluminum is evaporated onto the back side of the wafer. A boron glass spin-on solution is applied over the aluminum, followed by heating in O_2 at 950°C for two hours. The boron is carried into the silicon by the alumnum melt and lowers the sheet resistance of the back p$^+$ layer that freezes out when the wafer cools down to the silicon-aluminum eutectic temperature.

D. Gettering

Deleterious metal contaminants that serve as minority carrier recombination centers or low-resistivity paths are found in ribbon, cast, and low-quality Czochralski silicon. Among the most harmful are Fe, Ni, and Cu. Minority carrier lifetime is sensitive to Fe in p-type silicon at concentrations as low as 10^{10} cm^{-3}. At room temperature, iron can exist in the silicon lattice as an ionized interstitial atom (Fe$^+$) either singularly or paired to a boron acceptor (B$^-$). Several energy levels for the Fe$^+$B pair in the silicon bandgap have been measured[45]. One level is 0.10 eV above the valence band. Another is 0.55 eV below the conduction band. Heat treatment as low as 200°C for 10 minutes can cause

Fe⁺B to dissociate. When this happens, the interstitial Fe^+ produces an energy level at 0.43 eV above the valence band. Interstitial ionized Fe atoms are more effective recombination sites than Fe^+B pairs. Thus, any dissociation of the pairs will increase the bulk recombination rate and decrease the diffusion length. Recent results suggest this is the mechanism of the 2% to 5% photo-induced degradation seen in Czochralski silicon cells when they are initially exposed to sunlight[46]. Copper and nickel are fast diffusers and tend to preferentially segregate at the wafer surface where they form $NiSi_2$ and Cu_3Si which can cause shunts through the pn-junction[47]. The formation of contaminant silicides is sometimes seen as a haze on the wafer surface. During the gettering process, metal atoms are dissolved from a precipitated state and diffuse to the gettering site where they are stabilized (demobilized). Whereas a BSF reduces recombination velocity at the back surface, gettering of metal impurities in the bulk lowers dark current and improves performance by decreasing the rate of bulk recombination. Thus the effect of gettering is less dependent on the W/L ratio than is the effect of a back surface field. Interestingly, some interstitial ions increase minority carrier lifetime compared to the corresponding pair with boron. An example is CrB. The interstitial Cr^+ ion has a lower recombination activity than the CrB pair. However, the existence of the Cr^+ ion still degrades the lifetime compared to the case with no chromium.

In solar cell processing, gettering is always of the *extrinsic* or *external* type where metal impurities are drawn to the surface regions. This is contrasted to intrinsic gettering, used in the IC industry, where process-induced metal contaminants are moved away from the surface active region and into the nonactive bulk. Extrinsic gettering takes the form of either phosphorus in-diffusion from the front surface or aluminum alloying at the back surface. The phosphorus diffusion and aluminum alloying processes involved in routine pn-junction and BSF formation have built-in gettering effects. Two physical mechanisms for extrinsic gettering have been identified [48],[49],[50]: impurity segregation and injection of silicon self-interstitials. Segregation-induced gettering relies on enhanced solubility of the metal impurity in the gettering region at the gettering temperature. An example[51] is the backside gettering of a wafer by an Al-Si melt. The process is performed well above the eutectic temperature where the solubilities of various metals in the melt are orders of magnitude higher than in the silicon bulk. At 900°C, the solubilities of Cu, Ni, and Fe in intrinsic silicon are 4.4×10^{-4} at%, 1.5×10^{-4} at%, and 8.5×10^{-8} at%, respectively, as opposed to several atomic percent in the melt[52]. The large distribution coefficient provides the driving force for the segregation of the metal into the Al-Si melt. In the steady-state condition, most of the impurity moves into the melt where its concentration is limited only by the distribution coefficient between the two regions. During the cool-down, which is fast compared to the duration of the gettering, the impurity is trapped in the gettering region. Aluminum can be applied to the back side as a paste over a thermally grown oxide passivant and fired at 800°C [53]. While Al gettering occurs to some extend during the formation of a BSF, the optimum Al-gettering time is about 7 hours which is prohibitively long for commercial processes. By comparison, optimum gettering time for P in-diffusion is shorter (about 2 hours), but it is not as effective as the Al treatment. Aluminum gettering by rapid thermal processing, while not optimal, is desirable because it minimizes alteration of the front-side pn junction.

Gettering by P in-diffusion is only partially explained by solubility-enhanced impurity segregation. Most of the phosphorus gettering effect is accounted for by injection of silicon self-interstitials from the front surface into the bulk[54], i.e., the injection of silicon atoms that are forced to positions in between regular lattice sites as a result of phosphorus substitution. During phosphorus diffusion, the concentration of silicon self-interstitials is hundreds of times the equilibrium value. In thermal equilibrium, the concentration of metal atoms on substitutional sites M_s is much greater than the concentration of metal atoms on interstitial sites M_i. But the mobility of metal interstitials is much greater than the mobility of metal substitutionals. Consequently, the out-diffusion of metal atoms on substitutional sites occurs by the transformation of M_s into M_i and the outdiffusion of the metal interstitials. This involves the so-called *kick-out* mechanism for metal impurities that dissolve on both interstitial and substitutional sites. It relates the thermal equilibrium concentrations of metal impurity atoms and silicon self-interstitials I:

$$M_i \leftrightarrow M_s + I$$

The transformation process is self-limiting, by the above reaction, because it produces an undersaturation of I. Phosphorus in-diffusion is effective in gettering M_s because it relieves this limitation by injecting I and driving the equilibrium in the direction of greater M_i. Experiments have shown that metallic impurities gettered by P diffusion can become mobile by subsequent oxidations[55]. This is not seen when a Al layer is the gettering agent. For both Al or P gettering, it is more advantageous to integrate the gettering with the cell fabrication sequence rather than as additional steps. Each additional thermal step tends to lower the bulk diffusion length.

4.4 LIMITATIONS IMPOSED BY SCREEN PRINTING AND ENCAPSULATION

The usual accommodation with screen-printed commercial cells is to assure low emitter sheet resistance values (40-50 Ω /$\square$) and accept fairly deep junctions (0.5 to 0.8 μm) and poor blue response. This is the trade-off for inexpensive front-side metallization, high throughput, areal uniformity of the contact, and easy solderability. (Note that backside Al metallization does not solder easily. Some Ag areas are necessary for soldering the metal tabs that interconnect cells.) Low cost and high throughput are attributed to simple equipment needs and low chemical waste. The screen printing silver paste necessarily includes a very fine glass powder or frit to assure a strong mechanical bond between the metallization and the silicon surface. The paste is briefly fired at 700°C, and is often able to penetrate a passivation or anti-reflection film that might be present. Photolithography for opening windows in the dielectric is not necessary. For commercial terrestrial cells, this is an overwhelming advantage that screen printing has over evaporated metallizations. The technology has been ubiquitously used for these cells since the early 1970s.

There are disadvantageous with screen-printed metallizations other than poor blue response. It has a thick consistency when painted on the cell, but tends to run and liquify during the firing. This broadens the width of the line, and reduces the height. The aspect ratio and cross-sectional area are small, and this increases the series resistance that current sees as it passes through the fingers to the bus bar. The interfacial contact resistance is also high – over 5 mΩ cm^2. Increased resistance restricts the fill factor to about 0.77. Secondly, printed linewidths after firing are 150 to 200 μm. This increases the shading loss and decreases the short-circuit current. Metal coverage is typically 6% for the fingers and 4% for the bus strips that connect the fingers. Thirdly, the high surface concentration compromises surface passivation. However, recent results indicate the problems of silver paste screen printing can be ameliorated by processing enhancements suitable for a high-throughput belt-transport environment. Large area screen-printed cells with 15% encapsulated efficiency employing mechanical texturing, phosphorus gettering, and back-surface field formation have been demonstrated[56]. Other results[46] indicate screen printing can be improved to allow 80-μm production linewidths, contact resistance under 3 mΩ cm^2, metal sheet resistance under 3 mΩ /$\square$, and shadowing loss below 4%. This can be done on 100 Ω /$\square$ emitters. The lower series resistance and improved blue response boosts the efficiency of a 100-cm^2 gettered and passivated screen-printed single-crystal cell to over 17%. The cell in[46] has an Al BSF and a selective emitter structure that provided 100 Ω /$\square$ between the 60-μm contact fingers. While this is a heavily processed cell, it does not involve photolithography and it demonstrates that screen printed cells have much room for improvement. This includes faster and more effective formation of ohmic contacts. An example is the use of rapid thermal annealing (RTA) of the common spray deposited TiO$_2$ anti-reflection coating to yield a titanium silicide[57]. The silicide (developed in the silicon IC industry) lowers the interfacial contact resistance to as little as 10^{-7} Ω cm^2, thereby improving fill factor. The metallization is fired through the TiO$_2$ coating by tungsten halogen lamps that bring the surface temperature to over 1000°C for a few seconds, followed by rapid cooling. Good results are also obtained by a 700°C plateau for 120 seconds in argon ambient followed by a 10°C/s cool-down[58]. Interfacial contact resistance is sensitive to the cooling rate. Rapid thermal annealing is an increasingly popular trend both for its speed and its suitability for an in-line belt transport processing environment.

Poor blue response associated with screen printing has been relatively insignificant because of a characteristic of the encapsulation that removes much of the blue end of the spectrum before it reaches the cells. Low cost encapsulation uses 0.46-mm sheets of ethylene vinyl acetate (EVA) to hermetically seal (pot) the cells and their solder-coated copper connector ribbons to a rigid polymer backplane. Backplanes are often made of Tedlar (polyvinyl fluoride). The encapsulation process is rapid. For the new "fast-cure" grades of EVA, lamination of the five-layer cover glass/EVA/solar cells/EVA/Tedlar sandwich takes 12 minutes[59]. EVA costs about $4.75 per m². Other encapsulants (pottants) with suitable physical properties, e.g., Teflon (poly-tetra-fluoro-ethylene), are prohibitively expensive. The low cost, easy working properties, and thermo-mechanical stability of EVA make it an excellent encapsulant for terrestrial solar cells. However, over long periods of time in high insolation and high temperature environments, UV photo-oxidation and thermal oxidation processes tend to break down the EVA film. This induces the formation of acetic acid, to which is attributed a subsequent browning or yellowing of the film[60]. Both UV- and thermal-oxidation processes are associated with chain scission and crosslinking reactions. High concentrations of OH and C-O groups due to crosslinking in the presence of oxygen lead to yellowing[61]. Discoloration is also attributed to interactions of stabilizing and curing additives at elevated temperature and UV-B light (285-350 nm)[62]. Additionally, acetic acid can react with lead oxide in the glass frit of the screen printed metallization and form lead acetate. In this case, the ohmic contact between the metal finger and the silicon surface is lost. The extent of these effects is dependent on UV power density, module temperature, duration of exposure, and the specific EVA formulation. Sometimes discoloration has no effect on module efficiency[63], but in desert conditions with side mirrors providing light amplification early formulations showed significant degradation. To minimize the likelihood of discoloration for long-warranty products, module cover glass is formulated with high cerium oxide content to absorb the UV below 360 nm. Discoloration effects are increasingly rare, and the lead acetate problem is extremely rare, as new formulations of EVA become available. Discoloration rates are now about 40 times less than those seen in the early 1980s. Under most one-sun conditions, the new "fast-cure" formulations are expected to prevent long-term discoloration provided a UV-screening superstrate is in place[64]. Cerium oxide glass in now commonly used in the module sandwich, as opposed to the low-iron cover glass that was popular until about

1991. (Low-iron glass was used to maximize transmission of UV-B light.) It is the present necessity of a UV screen that lessens the importance of the poor blue response associated with screen printing. For commercial terrestrial modules, much of the blue end of the spectrum never reaches the cells anyway. Space cells do not have this problem as they require no encapsulation. As EVA formulations become more resistant to UV effects, the need for a UV screen will become relaxed, and more of the blue end of the spectrum will be allowed to reach the cells. Thus, developments in screen printing for better blue response will continue to be a major topic in commercial PV development.

After metallization and firing, cells are binned according to short-circuit current so they will be current-matched in the module. Solder-coated leads (tabs), several mm wide, are attached to screen-printed silver bond pads on the front and back surfaces (fig. 4.4-1).

Fig.4.4-1   Automated cell tabing process (*Photo courtesy Solarex*)

4.5 MODULE-SPECIFIC DEGRADATION MECHANISMS

For any photovoltaic technology where individual cells are combined to form a module, the module efficiency will tend to be about one or two percentage points *less than* the efficiency of the individual cells. This is not surprising when one considers several inevitable built-in factors that tend to degrade efficiency as cells are combined to form a module. These include series resistance, photocurrent mismatch, reflection from the module cover glass, and the presence of unproductive module area. The most significant of these factors is probably the *series resistance* that appears in the electrical connections between individual cells. Series connections are made by soldering metallic tape to the front-side metal contact of one cell and the back-side metal contact of another cell. By referring to the equivalent circuit model for a solar cell (Section 3.4), it is seen that a series connection between cells is merely a conducting path from the emitter of one cell to the base of the next cell. However, there is a resistance associated with the connecting metallic tape and the solder contact, and these add to the series resistance that is created during the manufacture of the cell when a metal contact is made to the semiconductor surface. Series resistance biases the cell's dark diode, and increases the magnitude of the dark current. By this mechanism, the short-circuit current of the cell is degraded. That is, as described in Chapter 3, the short-circuit current is less than the photocurrent, as some of the photocurrent is lost through dark current injection into the base of the cell.

Photocurrent mismatch is significant whenever cells are connected in *series*. As an example, consider three illuminated cells that are connected in series with a load to form a circuit. For purposes of illustration, assume the condition of very low series resistance and very high shunt resistance, and that the load is much greater than the series resistances of the individual cells. If the cells have identical I-V characteristics, then the voltage across the load will be the sum of the voltages produced by the individual cells, while the current through the load will be the same as the independent current of each cell. However, if the I-V characteristics of the three cells are not identical, then the performance of the series string is controlled by the poorest of the three cells. That is, the voltage across the load will still be the sum of the individual voltages, but the current through the load will be the smallest of the three individual currents that would be produced if the cells were operated independently of each other. During steady-state illumination, the requirement of continuity of current causes the current to be the same at all points in the circuit. The factors that limit the current in the lowest-current cell will *clamp* the current to the value seen in that cell. Solar cell manufacturers sort out their cells according to short-circuit current I_{sc}, and only use cells with similar values of I_{sc} in a given module. A manufacturer will typically bin the cells coming off the production line into twenty or more I_{sc} groups. However, there is always some current mismatch between cells, and this fact results in some degree of current clamping in each of the series strings that make up the module. The overall module efficiency suffers accordingly.

While the cells are encapsulated between sheets of plastic film (usually EVA) to assure hermiticity of the electrical contacts, a top cover glass is still necessary. Usually the glass is included in the EVA sandwich. The cover glass provides mechanical protection from colliding objects like hail stones. Glass composition is chosen for high transmissivity in the solar spectrum beyond 360 nm. Cover glasses also have anti-reflection coatings, but some *reflection* is always present. Reflection from the cover glass and plastic encapsulant attenuates the power density that reaches the solar cells, and degrades module efficiency.

Finally, *not all of the module's area* even produces a photocurrent. To avoid the accidental shorting of cells, it is necessary to leave a certain amount of space between the individual cells when they are laid out on the back plane of the module. There is also unproductive area taken up by the edge of the module's frame. Consequently, even in the ideal case where there is no series resistance in the intra-module connections, identical short-circuit currents, and no reflection from the cover glass, the module efficiency is still less than the efficiency of most of the individual cells.

The degrading effects of series resistance, current mismatch, reflection, and unproductive area, are intrinsic in the design of solar cells and the construction of the module. However, there also exists an extrinsic efficiency degrading effect caused by the *module's environment* rather than its engineering and material properties. This is the effect of partial shading of a module. It is *unique* to photovoltaics and has no analog in other power producing technologies. The partial shading of a module can be either transient, e.g., a falling leaf, or permanent, as in the case of severe soiling of the cover glass. While these may seem at first glance to be rather benign events, they can seriously impair the efficiency of a module. Additionally, if there is insufficient electrical protection, partial shading of a module can result in its permanent failure.

For a wide range of illumination intensities (not high concentration), the photocurrent from a cell is proportional to the intensity. When a cell is shaded from the sunlight, its current is drastically reduced. This produces a rather strange result for the case when the shaded cell is in a single string of cells. Depending on the configuration of the series strings in the module, and the configuration of the modules in the array, the partial shading of a single cell can produce various degradation and failure modes.

Consider, for example, the case of single-cell shading in a module with a single long string of n^+p cells. Because the shaded cell is connected in a *series configuration*, it will clamp the current in the string to a very low value. The current in a series connection is controlled by the poorest current producer. For a one-module system, the resulting small current in the circuit produces only a small voltage drop (current times resistance) across the load. If the shaded cell is near the bottom of the string, the p-side of the cell will be tied almost to ground, while the n-side of the cell will be strongly *reverse-biased* by the unshaded cells above it. If there are 36 cells in the string, each with V_{oc} of 0.6 V, and the bottom cell is shaded while the other cells are illuminated, then the shaded cell is reverse-biased by about 21 V. The current in the entire string is then clamped to the saturation current of the shaded cell's dark diode. For a module with only one string, the efficiency would fall essentially to *zero*. In an extreme situation, the pn-junction of the shaded cell could become permanently damaged by the high reverse bias. Thereafter, it behaves like a linear resistor, and dissipates the string's power in the form of heat. There are examples of entire modules being ruined by the shading of a single cell. In these rare situations, the plastic laminate is sometimes seared in the vicinity of the cell. Silicon diffused diodes can usually survive a reverse bias of at least 30 V and are not destroyed by shading. However, if several modules, each with one series string, are themselves connected in series, and one of the cells near the bottom of the stack becomes reverse-biased, the shaded cell and its module will likely be ruined. For this configuration, the entire module containing the shaded cell is reverse-biased. In principle, the performance of an entire array of modules can be severely degraded (if not ruined) by a single shaded cell.

Shading-induced degradation is completely reversible when the shading

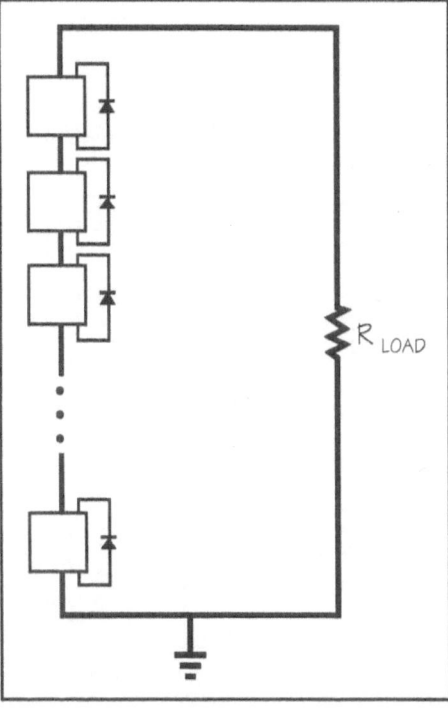

Fig. 4.5-1   By-pass diodes prevent reverse-bias of a module in a series string of modules.

event is terminated, provided the shaded cell has not been damaged. The degree of degradation can be restricted to a single module and the possibility of module damage can be avoided by use of by-pass diodes that effectively *isolate a module* that has become reverse-biased (fig. 4.5-1). Some manufacturers include by-pass diodes in the module junction box. For array voltages greater than 48 volts, when modules are connected in series, by-pass diodes are recommended[65]. If all the modules are in parallel, by-pass diodes are not required since the maximum reverse-bias on a module is limited to the forward bias of the other modules. However, for several parallel strings of modules, there is always the possibility of current from stronger strings being bled back to the system ground by way of the weaker strings. To prevent this and maximize power to the load, parallel strings of modules are isolated from each other by blocking diodes (fig. 4.5-2).

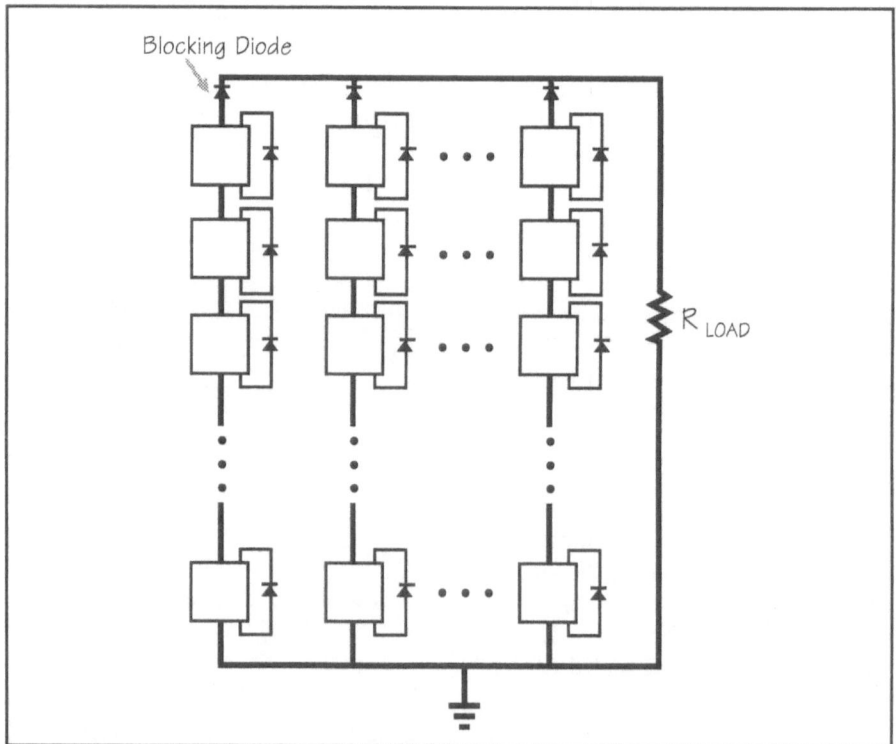

Fig. 4.5-2   Blocking diodes prevent current loss between parallel strings of modules.

4.6 HIGH-EFFICIENCY SPACE CELLS

The space cell market is driven by a need for high efficiency and dominated by thin (100-200 μm) silicon cells and GaAs cells on Ge substrates (fig. 4.6-1). While the

Fig.4.6-1   Panel consisting of GaAs/Ge cells for the Mars Pathfinder Lander.
(*Photo courtesy TECSTAR Inc. Applied Solar Division*)

specific power for most satellite PV *systems* is only about 15-25 W/kg, any savings in PV *system mass* translates into lower specific costs ($/kg) for the payload. Launch costs for low-earth orbit and geosynchronous orbit, not including payload and ground support expenses, are about $11 K/kg and $66 K/kg, respectively. Complete *end-of-life (EOL)* solar cell array costs are about $432 K/kW and $644 K/kW for Si and GaAs/Ge, respectively[66]. For every kilogram of photovoltaic equipment (cells, substrates, supporting masts, connectors, wiring, etc.) that is not needed on a satellite, roughly 0.6 kg of scientific or other instrumentation can be added to the vehicle. The mass specific power (W/kg) of the cell alone is not very significant. A much more significant parameter is power per unit area. This is because most of the mass of the PV system is in the form of *supporting structures* and associated equipment, and increases with the *area* of the cells. High efficiency cells minimize area needed for a particular power requirement. This lowers total PV system mass, thereby maximizing payload mass, and minimizing the specific cost for the payload ($/kg). Silicon space cells tend to be thin more for reasons of minimizing the volume subject to energetic-particle displacement damage than for reasons of mass. Merely making cells thinner has little effect on the specific cost of the array. Silicon, GaAs on Ge, and GaInP$_2$/GaAs/Ge tandem junction space-qualified cells have *beginning-of-life (BOL)* efficiencies of about 12.5 to 15%, 18.5%, and 21.5 to 24%, respectively. Their costs per square centimeter are approximately $2, $21, and $26 for thin-Si, GaAs/Ge, and GaInP$_2$/GaAs/Ge cells, respectively[67]. Another estimate establishes the ratio of prices between cells as x, (5 to 7) x, and (5 to 7) (1.15) x for Si, GaAs/Ge, and GaInP$_2$/GaAs/Ge, respectively[68]. As of 1996, GaInP$_2$/GaAs/Ge cells are only starting to enter commercial production. Their costs will eventually approach those of GaAs/Ge because both cells are made by the same process: metal-organic chemical vapor deposition. The tandem cell requires multiple layer growth, but this does not add significantly more expense in an automated MOCVD reactor process environment. However, the yield for the more complicated structure is lower than for GaAs/Ge. The tandem structure additionally requires reverse bias protection. At the cell level, both GaAs/Ge and tandem junctions are heavier than Si, and cost approximately five times as much per unit *power*. This is an acceptable trade-off for space cells because the higher areal specific power (W/m^2) for GaAs/Ge and tandem junction cells results in lower overall support structure mass and subsequent decreased cost per unit payload (fig. 4.6-2).

Since 1995, the strong trend has been toward GaAs on Ge cells. As of 1996, about 75% of new satellite power systems were based on non-silicon technology. By the year 2002, the dominant PV technology for new satellites will probably be the GaInP$_2$/GaAs/Ge tandem junction cell. The exception might be for intermediate altitude orbits where InP cells will show superior EOL performance after 10-year missions. Bulk InP is expensive, brittle, and heavy; development favors InP films on Si substrates. The p$^+$n polarity is easy to work with because Si is a donor dopant in III-V binary systems. However, the n$^+$p polarity performs better because $\rho_n \ll \rho_p$ in InP for a given doping level, and a lower emitter sheet resistance can be achieved. Eventually dual junction InP/InGaAs/Ge cells may become well-developed for use in intermediate altitude orbits where they will provide BOL and EOL performances above 20%.

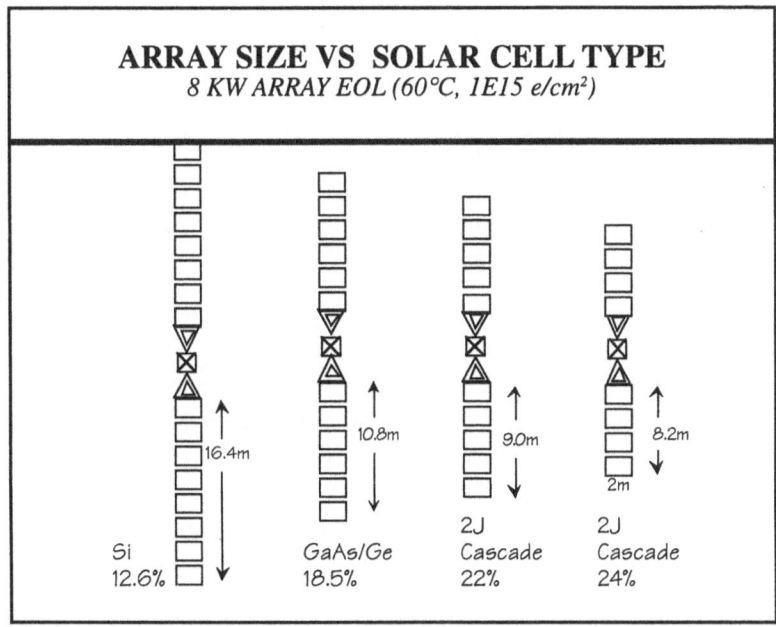

Fig.4.6-2   Comparison of array sizes for Si, GaAs/Ge, and GaInP$_2$ / GaAs / Ge technologies for a particular mission. (*Courtesy TECSTAR Inc. Applied Solar Division*)

For the Si, GaAs/Ge, and GaInP$_2$/GaAs/Ge tandem junction technologies now commercially available, a comparison of mass and cost requirements is provided in [69]. The comparison is based on the requirement of 2674 watts at EOL and a low-earth orbit altitude of 350 km. For these three technologies, the specific power for the *entire PV system* is about 14, 19, and 22.5 W/kg, respectively.

Table 5 shows that the total support mass savings that comes with high efficiency cells more than offsets the extra cost of these cells. Major cost improvements are afforded by new low-mass panel substrate materials on which the cells are mounted. GaAs/Ge cells (18.5%-efficient under AMO spectrum 28°C) mounted on graphite epoxy panels are commercially available and yield BOL panel specific power of 111 W/kg and panel areal power density (areal specific power) of 218 W/m^2. Panels made of similar material and employing GaInP$_2$/GaAs/Ge tandem cells (21.5%-efficient) are available with 112 W/kg and 254 W/m^2. These are compared to panels with 14.9%-efficient Si cells that yield only 178 W/m^2 BOL[70].

SOLAR ARRAY COMPARISON FOR LOW-EARTH ORBIT: Si, GaAs/Ge, AND TANDEM JUNCTION TECHNOLOGIES

		Si	GaAs/Ge	Tandem
Mass of cells, wiring, etc.	(kg)	48	47	38
Mass of PV system	(kg)	192	141	119
Payload mass above Si baseline	(kg)	–	31	45
Cost of cells, wiring, etc.	($M)	2.5	4.2	3.6
Cost of total PV system	($M)	3.3	4.7	4.0
Cost of payload launch	($K/kg)	552	90	58

Table 5   Solar array comparison for low-earth orbit for Si, GaAs/Ge, and tandem junction technologies. P_{EOL} = 2674 W. (*Adapted from ref. 69*)

In general, space cells are unencapsulated, but require cover glasses to reduce operating temperatures and to attenuate particle damage. Metallizations and cover glasses are thermo-mechanically matched to the semiconductor to accommodate temperature changes from radiative heating and cooling as the photovoltaic panel moves in and out of the sunlight. Panels are qualified to several thousand or more thermal cycles between roughly +/-100°C. Cover glasses are attached to the cells with transparent silicone adhesive. Some covers have a thin film of conductive transparent coating in the form of evaporated indium tin oxide (ITO) which allows accumulated charge from particle fields to be bled away to prevent destructive electrostatic discharge[71]. The cells with cover glasses are wired together and cemented to a polymer film which is electrically insulating but thin enough to be thermally conductive. A 50-μm Kapton (polyimide) film is a common choice because of its dimensional stability at over 200°C. For a body-mounted array, the cell/polymer sandwich is cemented to an aluminum facesheet which is then interfaced to an aluminum honeycomb structure for radiative heat dissipation. Alternatively, polymer and cells are cemented to a flexible carbon fiber facesheet and attached to an aluminum honeycomb to form a panel on a deployable mast. In some designs, a flexible substrate is used that can be rolled out like a carpet.

Cells intended for low-earth or geosynchronous orbits use 100- to 200-μm cover glasses. Intermediate orbits where more energetic particles (particularly protons) reside require upwards of 600-μm covers. This thickness proves too opaque for some missions. However, the next generation of communications satellites (which strives to achieve long-term whole-earth coverage with only a few satellites) is intended for these orbits. Thus, improved less-opaque covers are under development[72]. Cover glasses have been used which are as thin as 80 μm and have an anti-reflective coating on the front surface and also a far-ultraviolet reflective coating. Cells using Ge substrates have been deployed with covers having an IR reflective coating on the back surface[73]. The IR reflective coating slightly lowers short-circuit current, but prevents severe degradation of the open-circuit voltage from heating of the cell. In-orbit cell temperatures are 14 to 20 K less than with conventional AR/ITO cover glasses. This boosts in-orbit cell efficiency as much as 6% (relative)[74]. In general, to minimize heating, it is desirable to reflect the incident light that is not in the response band of the solar cell. For Si cells, the response band is 350-1150 nm; for GaAs cells, the band is 350-900 nm. ITO coatings are sufficiently thin that they do not affect the thermo-optical properties of the cover glass and can be combined with anti-reflection, far-UV reflecting, and infrared reflecting coatings[75].

Energetic particles can also enter the cell from the back side in the case of a panel on a deployed mast. However, a typical 120-μm carbon fiber facesheet and a 22-mm honeycomb afford shielding equivalent to a 350-μm backside cover glass[76]. Table 6 shows estimated values for 5-year EOL panel-level areal and mass specific powers for several technologies on lightweight deployable arrays in low-earth orbit (LEO). This includes high-performance Si cells, GaAs cells grown by MOCVD on bulk GaAs, bulk InP, and also ultra-thin (10 μm) GaAs cells that show high resistance to particle damage. As of 1996, ultra-thin GaAs and InP cells were still in the development stage.

ESTIMATED DEPLOYABLE-ARRAY AREAL AND MASS SPECIFIC POWERS
Five-Year EOL, LEO, 50° C

	Si	InP	GaAs/GaAs	GaAs/Ge	UTC GaAs
W/m²	145	200	225	205	230
W/kg	55	55	63	74	105

Table 6   Estimated deployable-array areal and mass specific powers for 5-year EOL in LEO at 50°C. (*After ref. 76*)

Space cells are designed with smaller area than terrestrial cells, typically 16 to 36 cm². High-efficiency float-zone Si cells are necessarily small due to ingot cross section. The MOCVD growth of GaAs/bulk GaAs, GaAs/Ge, and tandem junction cells is limited to about 80 cm² substrates to achieve acceptable layer uniformity[77]. Additionally, cells are attached to surfaces that are either not flat or subject to mechanical vibration, and small size minimizes the liklihood of cracking (fig. 4.6-3).

Fig.4.6-3   GaAs/Ge cells affixed to a section of a spacecraft. (*Photo courtesy TECSTAR Inc. Applied Solar Division*)

Silicon space cells require shallow junctions (0.15 μm) and evaporated metallizations (TiPdAg front, and AlTiPdAg back)[78]. Aluminum serves as a back surface optical reflector so that the effective path length of near-infrared light is increased. More importantly, because the cell is attached to a large opaque mass, it is important that all wavelengths beyond the cutoff wavelength are reflected out of the top side of the cell, thereby reducing cell temperature. The back surface field is made either by alloying the aluminum or by a separate boron diffusion. "Wrap-through" designs (fig. 4.6-4) are popular with Si space cells because the easy panel assembly lowers overall cost by about 25%.

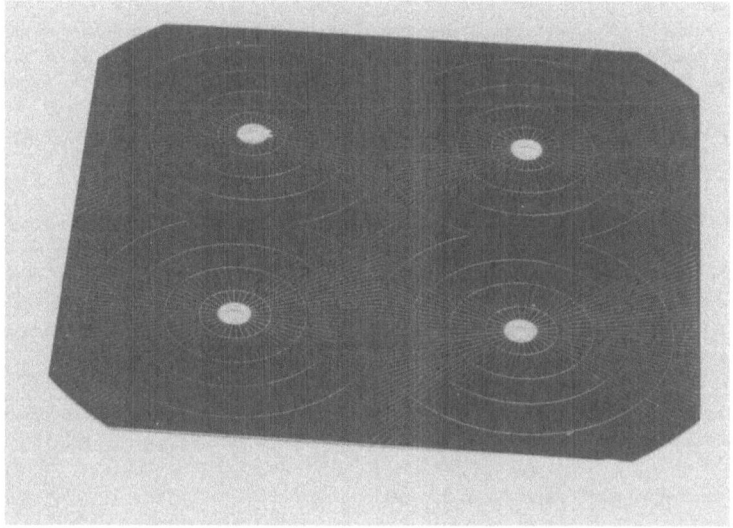

Fig.4.6-4   Silicon cell with wrap-through design. (*Photo courtesy TECSTAR Inc. Applied Solar Division*)

In the wrap-through approach, the front TiPdAg metallization is brought through a hole to the back side of the cell. This allows both emitter and base contacts to be made on the back side. Several holes (3-5 mm) are cut through the wafer by laser. A dielectric layer is sputtered around the vicinity of the hole for electrical isolation before depositing evaporated metal (100 nm Ti/50 nm Pd/several µm Ag). The low front surface obscuration increases J_{sc}. There is a similar "wrap-around" design that deposits dielectric on the edges of the cell before front-side metallization so that front contact can be made from the backside.

REFERENCES

[1]   M.A. Green, et al., "Solar cell efficiency tables (version 7)," *Progress in Photovoltaics*, pp. 59-62, Jan-Feb, 1996.

[2]   B.E. Yoldas, "Diffusion of dopants from optical coatings and single step formation of antireflective coating and p-n junction in photovoltaic cells," *J. Electrochem. Soc.*, vol. 127, pp. 2478-2481, Nov. 1980.

[3]   J.C.C. Tsai, "Shallow phosphorus diffusion profiles in silicon," *Proc. of the IEEE*, vol. 57, pp. 1499-1506, Sept 1969.

[4]   M.Y. Ghannam, et al., "636 mV open circuit voltage multicrystalline silicon solar cells on Polix material: trade off between short circuit current and open circuit voltage," *Conf. Rec. of the 23rd IEEE Photovoltaic Specialists Conf.*, pp. 106-110, May 1993.

[5]   M.D. Rosenblum and J.I. Hanoka, U.S. Patent 5,270,248, "Method for forming diffusion junctions in solar cell substrates," Dec. 14, 1993.

[6]   N. Mardesich, "Solar cell efficiency enhancement by junction etching and conductive AR coating processes," *Conf. Rec. of the 15th IEEE Photovoltaic Specialists Conf.*, pp. 446-449, May 1981.

[7]   Private communication, A. Rohatgi, Oct. 1995.

[8]   H.E. Elgamel, et al., "Industrial screen printing solar cell process using selctive emitter and hydrogen passivation," *Proc. of the 11th E.C. Photovoltaic Solar Energy Conf.*, pp. 389-392, Oct. 1992.

[9]   H.S. Carslaw and J.C. Jaeger, *Conduction of Heat in Solids*, 2nd ed., Oxford University Press, New York, 1959.

[10] Notes from panel discussion on implementation of advanced processing, *Fifth Workshop on the Role of Impurities and Defects in Silicon Device Processing*, Copper Mountain, CO, Aug. 1995.

[11] M.A. Green, et al., "Enhanced light trapping in 21.5% efficient thin silicon solar cells," *Proc. of the 13th European Photovoltaic Solar Energy Conf.* pp. 13-16, Oct.1995.

[12] T.Tiedje, et al., "Limiting efficiency of silicon solar cells," *IEEE Trans. on Electron Devices*, vol. ED-31, pp. 711-716, 1984.

[13] H.J. Hovel, *Semiconductors and Semimetals, vol. 11*, ed. by R.K. Willardson and A.C. Beer, Academic Press, New York, pp. 95-98, 1975.

[14] J. Zhao, et al., "21.5% efficient 47-μm thin-layer silicon cell," *Proc. of the 13th European Photovoltaic Solar Energy Conf.*, pp. 1566-1569, Oct. 1995.

[45] J.P. Kalejs, "Study of Fe and Cr pairing with B in polycrystalline silicon," *Proc. of the 12th E.C. Photovoltaic Solar Energy Conf.*, pp. 52-55, Apr. 1994.

[46] J.H. Reiss, R.R. King, and K.W. Mitchell, "Degradation of bulk diffusion length in CZ silicon solar cells," *Extended Abstracts of the Fifth Workshop on the Role of Impurities and Defects in Silicon Device Processing*, pp. 120-123, Copper Mountain, CO, Aug. 1995.

[47] D. Gilles and H. Ewe, "Gettering phenomena in silicon," *Semiconductor Silicon/1994 - Proc. of the 7th Int. Symp. on Silicon Material Science and Tech.*, pp. 772-783, 1994.

[48] T.Y. Tan, R. Gafiteanu, and U.M. Gosele, "Toward understanding and modeling of impurity gettering in silicon," *Extended Abstracts of the Fifth Workshop on the Role of Impurities and Defects in Silicon Device Processing*, pp. 93-100, Copper Mountain, CO, Aug. 1995.

[49] W. Schröter, M. Seibt, and D. Gilles, *Materials Science and Technology: A comprehensive Treatment*, ed. by W. Schroter, vol. 4, pp. 576-586, VCH, New York, 1991.

[50] W. Schröter, E. Spiecker, and M. Apel, "Gettering of metal impurities in silicon," *Extended Abstracts of the Fifth Workshop on the Role of Impurities and Defects in Silicon Device Processing*, pp. 85-92, Copper Mountain, CO, Aug. 1995.

[51] L.A. Verhoef, et al., "Gettering in polycrystalline silicon solar cells," *Material Science Engineering*, vol. B7, pp. 49-62, 1990.

[52] E.R. Weber, *J. Appl. Phys.*, vol. A30, p. 1, (1983).

[53] J. Nijs, et al., "Latest efficiency results with the screen printing technology and comparison with the buried contact structure," *Conf. Rec. of the First World Conf. on Photovoltaic Energy Conversion*, pp. 1242-1249, Dec. 1994.

[54] W. Schröter and R. Kühnapfel, *"Model describing phosphorus diffusion gettering of transition elements in silicon,"* *Appl. Phys. Lett.*, vol. 56, p. 2207-2209, May 1990.

[55] J.M. Gee, data presented at *Crystalline Silicon Research Cooperative Review*, Crystal City, VA, Jan. 1996

[56] S. Narayanan, et al., "15% efficient 230 cm^2 screen printed multicrystalline silicon cell process development," *Abstracts of the 25th IEEE Photovoltaic Specialists Conf.*, no. 141, May 1996.

[57] J.P. Boyeaux, et al., "Towards an improvement of screen printed contacts in multicrystalline silicon solar cells," *Proc. of the 11th E.C. Photovoltaic Solar Energy Conf.*, pp. 503-506, Oct. 1992.

[58] H. El Omari, J.P. Boyeaux, and A. Laugier "Screen printed contacts formation by rapid thermal annealing in multicrystalline silicon solar cells, " *Abstracts of the 25th IEEE Photovoltaic Specialists Conf.*, no. 136, May 1996.

[59]   Solar cell encapsulant product literature from Springborn Materials Science Corp., Enfield, CT, May 1996.

[60]   E. Cuddihy, et al., "Module encapsulation," vol. 7 in *Flat-Plate Solar Array Project Final Report*, JPL Publication 86-31, pp. 49-52, Oct. 1986.

[61]   Z. Lin and W.B. Berry, "FTIR evaluation of UV and thermally degraded ethylene vinyl actate (EVA) solar module encapsulant," *Proc. of the 11th E.C. Photovoltaic Solar Energy Conf.*, pp. 1094-1097, Oct. 1992.

[62]   J.P. Galica, et al., "Advanced development of non-discoloring EVA-based PV encapsulants," *Proc. of the 13th E.C. Photovoltaic Solar Energy Conf.*, pp. 2370-2372, Oct. 1995.

[63]   R.C. Petersen and J.H. Wohlgemuth, "Stability of EVA modules," *Conf. Rec. of the 22nd IEEE Photovoltaic Specialists Conf.*, pp. 562-565, Oct. 1991.

[64]   W. Holley, et al., "Development of nondiscoloring EVA-based encapsulants," *Abstracts of the 25th IEEE Photovoltaic Specialists Conf.*, May 1996.

[65]   *Stand-Alone Photovoltaic Systems*, Sandia National Laboratories publication SAND87-7023, pp. 30-32, revised edition of Nov. 1991.

[66]   E.L. Ralph, "High efficiency solar cell arrays system trade-offs," *Conf. Rec. of the First World Conf. on Photovoltaic Energy Conversion*, pp. 1998-2001, Dec. 1994.

[67]   E.M. Gaddy, "Cost performance of multi-junction, gallium arsenide, and silicon solar cells on spacecraft," preprint of paper for 25th IEEE Photovoltaic Specialists Conf., May 1996.

[68]   Private communication. F. Ho, June 1996.

[69]   E.M. Gaddy, "Cost trade between multijunction, gallium arsenide and silicon solar cells," *Progress in Photovoltaics*, vol. 4, pp. 155-161, Mar-Apr, 1996.

[70]   Product literature, Tecstar, City of Industry, CA, May 1996.

[71]   C.A. Kitchen, et al, "Teflon bonding of solar cell assemblies using Pilkington CMZ & CMG coverglasses - now a production process," *Conf. Rec. of the First World Conf. on Photovoltaic Energy Conversion*, pp. 2058-2061, Dec. 1994.

[72]   S.J. Morris, et al., "Solar cell cover glasses for the intermediate circular orbit type of spacecraft," *Abstracts of the 25th IEEE Photovoltaic Specialists Conf.*, no. 68, May 1996.

[73]   M.R. Brown, et al., "Characterization testing of Measat GaAs/Ge solar cell assemblies," *Progress in Photovoltaics*, vol. 4, pp. 129-138, Mar-Apr 1996.

[74]   G. Jones, et al., "The infra red reflecting coverglass for silicon and GaAs solar cells used in near earth

& geostationary orbits," *Conf. Rec. of the First World Conf. on Photovoltaic Energy Conversion*, pp. 2054-2057, Dec. 1994.

[75]   R. Herschitz and A. Bogorad, "Space environmental testing of blue red reflecting coverglasses for gallium arsenide and high efficiency silicon solar cells," *Conf. Rec. of the First World Conf. on Photovoltaic Energy Conversion*, pp. 2189-2191, Dec. 1994.

[76]   T.A. Cross, R. Kimber, and C. Goodbody, " Solar panels for microsatellites: GaAs, GaAs/Ge, and beyond," *Conf. Rec. of the First World Conf. on Photovoltaic Energy Conversion*, pp. 1970-1977, Dec. 1994.

[77]   Private communication, P. Iles, June 1996.

[78]   Product literature, Spectrolab Inc., Sylmar, CA, May 1996.

5

NON-INGOT AND NOVEL
TECHNOLOGIES

In 1995, 85% of the worldwide terrestrial module production of 81 MW consisted of three generic crystalline silicon technologies[1]: single-crystal (57%), cast polycrystalline (25%), and polycrystalline ribbons (3%). There was also a very small amount of commercial PV made from silicon sheet through recrystallization of silicon powder on inexpensive substrate. Thus, 82% of commercial module production was done with Si *ingot-based* cells. It is clear that crystalline silicon, both ingot-and non-ingot-based, will remain the dominant form of PV until at least the year 2001. Primary applications will remain off-grid electronics, lighting, home electrification, and water pumping. This is consistent with the observation that about 70% of U.S. production in 1995 was exported, and much of that was for applications in developing countries. The longer term hope for PV, though, is for domestic use in the form of competitive large-scale *central-station power generation.* As of 1996, the preponderance of opinion was that this optimistic long-term photovoltaic goal will not be met with ingot-based silicon technology. Instead, amorphous silicon modules[2] and non-silicon thin-film cells and modules are seen as more likely candidates. Thin-film modules based on amorphous silicon, polycrystalline cadmium telluride, or polycrystalline copper indium diselenide alloys are suitable for high-throughput semi-continuous production by either chemical vapor or physical vapor deposition. The substrate can be polymer or stainless steel on a reel-to-reel roll, or an individual glass plate. This approach blurs the distinction between cells and modules and allows *monolithic integration* of the entire process. The individual wafer-level tasks of material separation, sorting of cells, cell layout, and electrical interconnection are eliminated. Cells are differentiated during the module formation process by scribing the films. Monolithic integration promises the inexpensive production environment that will be necessary for competitive central-station photovoltaic power generation.

These relatively new PV technologies have already achieved very impressive conversion efficiencies. As of May 1996[3], experimental cells with areas on the order of one square centimeter have reached 1-sun total area efficiencies of 14.2% for cadmium telluride and 17.7% for copper indium gallium diselenide (CIGS) . The high values for prototype *large-area* monolithically integrated modules on glass substrates are even more impressive: 9.1% for a 6728 cm² CdTe module and 11.2% for a 3830 cm² CIGS module. These technologies are being aggressively developed[4]. However, CdTe and CIGS thin films have significant manufacturing reliability and process yield problems and have not been made in large commercial volumes. And neither technology has yet approached the above module performances with flexible substrates. Amorphous silicon modules are relatively more mature from a commercial standpoint. They are now made in large volume – 13.5% of worldwide shipments in 1995 – more than half of which is for "indoor PV" applications such as power sources for hand-held calculators and cameras. Modules are made on both glass and flexible substrates, and have been integrated into building materials such as roofing shingles. But the mechanism of the Staebler-Wronski effect[5], whereby amorphous silicon cells suffer irreversible efficiency degradation during the first several hundred hours of illumination, is still a topic of intense investigation[6]. Commercial amorphous silicon modules have *stable* efficiencies of only 6% to 9%.

Meanwhile, for the near- and intermediate-terms, some of the inherent expenses in ingot-based technology, particularly wafer sawing and kerf loss, favor increased commercialization of silicon *ribbons*. As previously discussed, ribbons have problems associated with their tendency for high concentrations of impurity contamination and thermal-stress induced defects. These problems are being resolved, as indicated by presently available successful commercial products, while the disadvantages of ingot-based material are more limited in their remediation. Ribbons show great promise for manufacturing cost reduction (material and labor) to values well under $2/W_p$ by the year 2000.

Besides inexpensive amorphous Si, CdTe, and CIGS films, large-scale central-station PV generation might also take the form of gallium arsenide-based thin films on germanium substrates, and passively-cooled concentrator systems. The latter tend to use either high-efficiency Si cells under medium concentration (< 100 X) or GaAs-based cells under very high concentration (500 to 1000 X). Both crystalline Si and GaAs cells become more efficient under concentrated light, provided they are adequately cooled[7]. Gallium arsenide is particularly well suited for concentrator applications because of its high efficiency and tolerance for elevated temperatures compared to silicon.

A photovoltaic development related to the issues of large-scale use (if not central-station use) and high concentration is thermophotovoltaics (TPV). Large-scale use could be realized by low-NO_x emission power systems for electric vehicles and co-generation of heat and electric power for buildings and military applications. Thermophotovoltaics is an interesting connection between fossil fuels and photovoltaics. It is also being developed for deep-space power systems where heat is provided by radioisotope sources. All TPV systems are inherently concentrator systems because of the proximity of the cells to the thermal emitter. Consequently, high-efficiency crystalline cells (Si, GaSb, or GaAs-based, including ternary alloys) are always chosen for the photovoltaic part of the system. Thermophotovoltaics will create increased demand for high-efficiency silicon cells that are not cost-effective for flat-plate terrestrial photovoltaics.

5.1 NON-INGOT CRYSTALLINE SILICON CELLS

Non-ingot crystalline silicon cells exist either as solution-grown thin films (< 50 μm) on crystalline silicon substrates[8], as recrystallized or solution-grown films (100 to 300 μm) on seeded expansion-matched ceramic substrates[9], or as conventionally-thick (250 to 300 μm) ribbons shaped from a molten state[10]. Unlike ribbons, thin-film *crystalline* silicon cells are not yet commercialized. Thin-film crystalline silicon is complicated by the need for a crystalline substrate or seed on which fairly large grains (0.1 mm) can form, and by the need for effective light trapping. For ribbons, performance tends to be compromised by dissolution of the shaping die or shaping surface into the meniscus of the melt, and by defects induced by the huge cooling gradient inherent in ribbon processes. There are trade-offs for these difficulties. The trade-off for thin-films is that they will allow great tolerance for poor diffusion length. In very thin film silicon cells[11], photogenerated carriers would only have to travel several microns to be collected. Effective anti-reflection coatings and light trapping would guarantee almost 100% transformation of usable light into collected current. Hydrogen passivation of thin films would be more effective than in bulk material because the hydrogen could penetrate the entire thickness of the cell. For ribbons, the trade-off is that ingot sawing is avoided and, for processes that have a large-area meniscus, the areal throughput per growing machine can be much greater than for ingot-based silicon. Over 20 ribbon technologies have been developed. This includes methods for forming silicon sheets by crystallizing silicon powder with a laser or other heat source[12]. Only three ribbon technologies (EFG, dendritic web, and edge-supported pulling) have reached pilot production.

Thin-film crystalline silicon cells have been made by liquid phase epitaxy in which Si is deposited on a crystalline substrate from a metal solvent (e.g., Cu, Sn, Ga, or Al). A charge of metal in a quartz crucible with an argon ambient is saturated with Si at some temperature above the eutectic temperature. The substrate is dipped in the liquid. As the temperature is slowly lowered, silicon crystallizes on the substrate with the same crystal orientation as the substrate. At the initiation of the process, temperature is maintained so that the native oxide and surface of the substrate are removed by dissolution into the metal/Si solution. This assures a very clean surface on which growth begins. The relatively high solvent power of Cu (0.05 wt.% Si per degree C near 940°C) and the low solid solubility of Cu in Si, make Cu the preferred solvent. By lowering the temperature very slowly so that the growth rate is under 1 μm/min, high-quality films free of Cu/Si inclusions are obtained. Inclusions are to be avoided as they tend to short the junction. For Cu/Si solutions, the process can be started at about 950°C. The low temperature compared to that of ingot growth (1430°C) minimizes the transport of carbon impurity from the graphite susceptor that holds the crucible. Cells formed by phosphorus diffusion into 5-μm thick p-type films on Czochralski substrates have diffusion lengths over 100 μm, and efficiency over 15%. The performance is slightly less than that of similarly diffused CZ control cells and has no commercial advantage over conventional cells. However, transfer of the process to metallurigical-grade substrates would be desirable from a cost standpoint. So far, LPE thin film cells on MG substrates have yielded efficiencies less than half that of similarly processed conventional cells[13].

Other studies have formed polycrystalline films on foreign substrates by crystallizing an amorphous film through annealing, or by depositing high-quality CVD polysilicon on small-grain polycrystalline material. Small-grain p-type crystalline silicon films have been formed by capping hydrogenated amorphous silicon films with aluminum, followed by vacuum annealing at 500°C. The low temperature anneal crystallizes the material while the aluminum layer traps hydrogen which passivates the grain boundaries. Aluminum simultaneously dopes the material. While the grains are only a few microns wide, they are smaller than the diffusion length. This approach[14] could have application in multijunction thin-film cells.

Ribbon technologies rely on the solidification of a melt in the form of a sheet, rather than an ingot. All ribbons solidify from a meniscus where molten silicon contacts the shaping element. Meniscus geometry and pull rate determine the rate and direction of cooling and the subsequent grain morphology. Ribbons can be classified by their meniscus type. In the first type, the meniscus is very short, on the order of the ribbon thickness. An example is EFG ribbon (fig. 5.1-1). In EFG, a melt rises vertically through a graphite die by capillary action. The die shapes the ribbon. After the initiation of the process, the ribbon is mechanically pulled upward at about 20 mm/min. For this type of meniscus, the melt shares a large surface area with the shaping element. This tends to create thermal and mechanical instabilities that affect the grain morphology. Additionally, the shaping element tends to dissolve and contaminate the solidifying ribbon with carbon and SiC precipitates[15]. As a die dissolves, the width and thickness of the ribbon slowly increase. Eventually, the die has to be replaced. Another example is the method of shaping by a substrate where the substrate is pulled vertically through the melt from a slot at the bottom of the crucible.

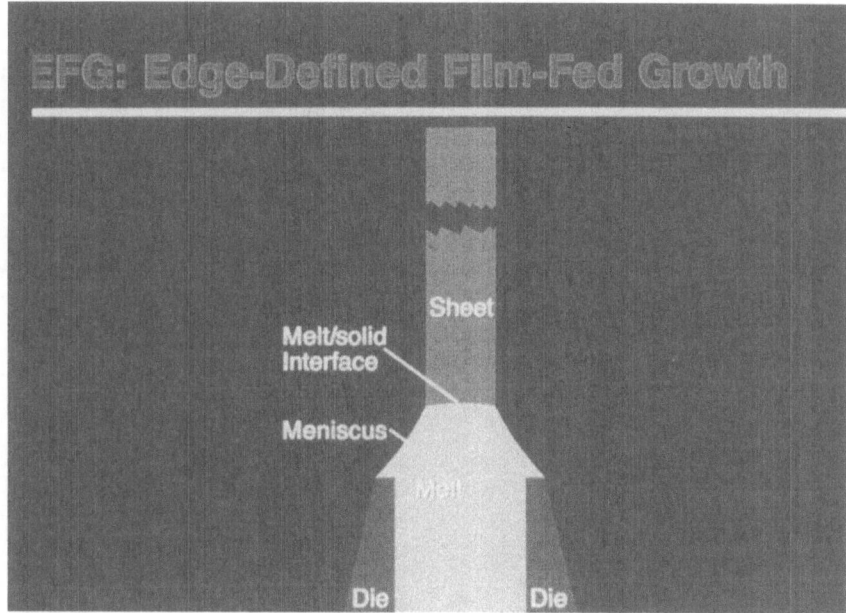

Fig. 5.1-1   Schematic of EFG ribbon emerging from a die. (*Photo courtesy ASE Americas, Inc.*)

The second group of ribbons have a free-rising meniscus where the molten silicon makes little contact with the shaping element. Examples are dendritic web and edge-supported pulling (ESP or string ribbon). In dendritic web, the meniscus is pulled vertically out of a melt guided by silicon dendrites positioned at the surface of the melt. A silicon button for pulling is grown from a dendrite and attached to the top of the meniscus at the beginning of the process. No foreign material touches the meniscus. Ribbon growth proceeds as the two guide dendrites propagate single-crystal growth from each edge of the solidifying meniscus. The result is a bicrystal ribbon with twin boundaries at the center where the two monocrystals meets. Edge-supported pulling[16] incorporates a pair of parallel graphite or quartz filaments that pass vertically through the melt from a pair of holes in the bottom of the crucible. Silicon surface tension prevents molten silicon from leaking through the holes. As the parallel filaments are continuously pulled out of the melt at the top surface, the suspended meniscus solidifies into a multicrystalline ribbon that is initiated by a seed. The filaments permanently remain in the ribbon at the edges. ESP tends to be more controllable than EFG and dendritic web because of its relative insensitivity to temperature fluctuations at the meniscus. For dendritic web and ESP, grain growth is parallel to the ribbon surface

The third group of ribbons is characterized by a meniscus with a large solid/liquid interface area compared to Wt, where W is the width and t is the thickness of the ribbon. Examples are horizontally grown ribbons where a large-area meniscus forms on a substrate passing over the surface of the melt. This method has a very fast growth rate as the latent heat of fusion is quickly removed through the large interface area. Grain growth is orthogonal to the surface of the ribbon. For most of the ribbon methods in these three meniscus categories, reasonable cell efficiencies (11% to 14%) have

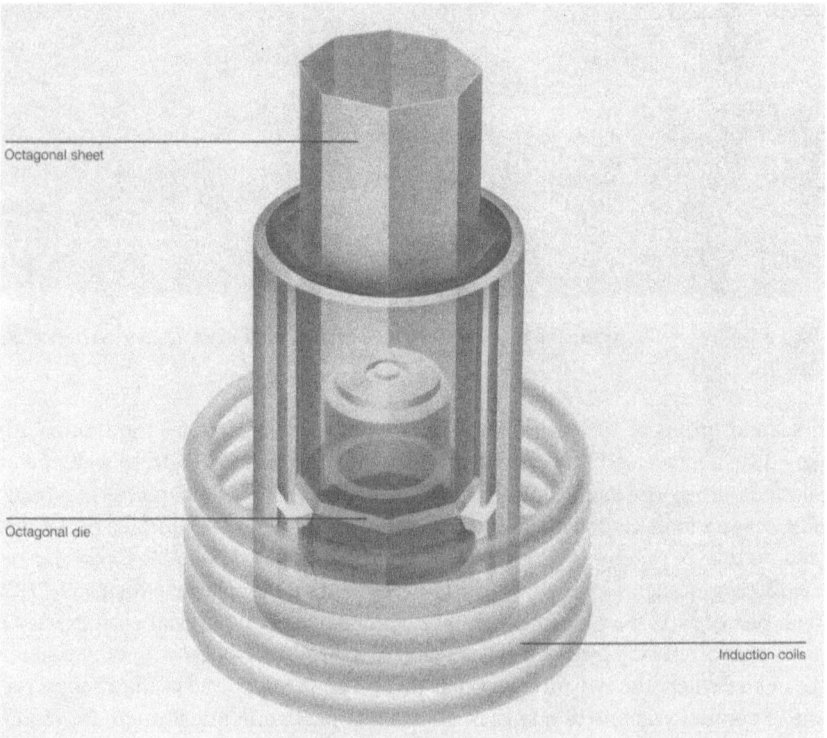

Fig. 5.1-2   In EFG, molten silicon rises through a slot in the die to form an octagon tube. (*Courtesy ASE Americas, Inc.*)

been demonstrated on a laboratory scale[17]. However, as of 1996, only the EFG method has demonstrated a commercially viable ribbon cell.

In the commercial EFG process, a meniscus is drawn through an eight-sided graphite shaping die 10 cm on a side to produce a closed eight-sided 300-μm thick ribbon (fig. 5.1-2). Growth is initiated in an argon ambient by point seeds along the width of the die so that the meniscus extends along all eight sides. The EFG material is predominantly

Fig. 5.1-3
Several
computer-
controlled EFG
ribbon pulling
machines.
(*Photo courtesy
ASE Americas,
Inc.*)

<110> surface orientation, with a high density of electrically inactive twin boundaries, and a low density of high-angle grain boundaries. Because of heat piping in the tube, grain growth is neither parallel nor orthogonal to the surface of the ribbon. When the tube reaches about 4.6 m in length, it is disengaged from the die-crucible assembly and is ready for slicing into individual flat pieces (fig. 5.1-3). An interesting characteristic of this and some other ribbon methods is that the crucible can be continuously re-charged, so that a huge amount of ribbon can be grown from a single graphite crucible. While there is a practical limit to the length of an octagon tube, one tube can be disengaged and another tube started without shutting down the crucible. This has been demonstrated by the growth of 100 full length tubes (equivalent to 350 m² of silicon sheet or 35,000 cells) from a single crucible (fig. 5.1-4).

Fig. 5.1-4   Section of an EFG octagon, 10 cm on a side, and finished solar cells.
(*Courtesy ASE Americas, Inc.*)

Wafer separation is accomplished by an infrared laser. While silicon is transparent in the infrared at usual sunlight concentrations, the laser intensity produces highly energetic electrons in the conduction band (hot electrons) which temporarily change the refractive index of the silicon so that it becomes an absorber. The silicon then melts along the laser scribe lines and the entire tube is separated into 10-cm square pieces (fig. 5.1-5) ready for processing with only minimal surface cleaning. Overall material loss in the disaggregation of the octagon tube into wafers is about 8%. Except for slight bowing on the edges at the octagon vertices, the pieces are flat and ready for processing into cells. The EFG process is highly developed and consistently produces cells with 14% efficiency under 1-sun AM1.5 spectrum.

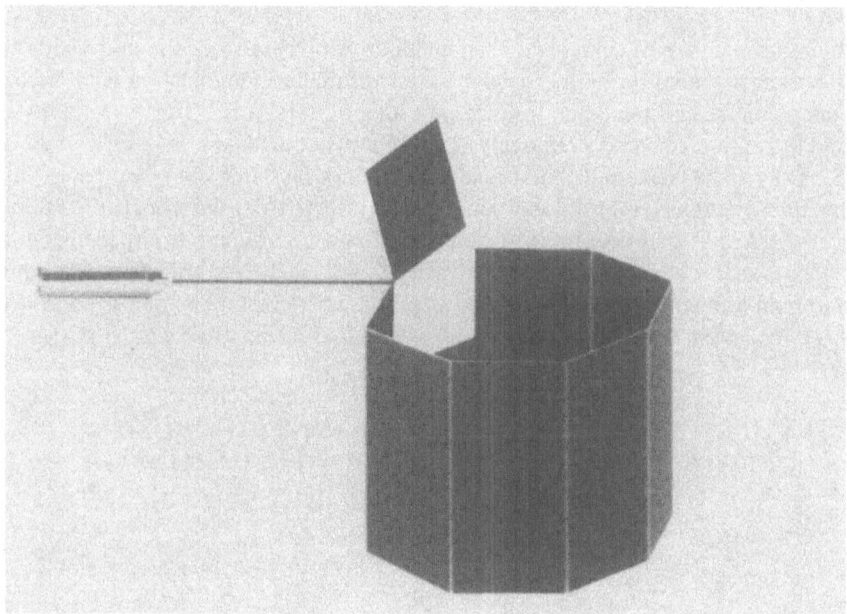

Fig. 5.1-5   Separation of an EFG octagon into 10-cm square pieces by laser scribing. (*Photo courtesy ASE Americas, Inc.*)

5.2 III-V MATERIAL-BASED SYSTEMS

III-V material systems include gallium arsenide, indium phosphide, and the ternary alloys of these compounds. GaAs and InP are direct-gap materials, but their ternary alloys can be direct or indirect, depending on the mole fraction of the third element. The mole fraction also affects the magnitude of the bandgap. With direct bandgaps, III-Vs have short absorption lengths and thin films are sufficient to collect the usable part of the spectrum. Bulk growth of III-V crystals is problematic because the constituent elements have different vapor pressures and careful control of the growth atmosphere is required in order to maintain stoichiometry. III-V ingots are much smaller than Si ingots. Additionally, III-V bulk crystalline blanks tend to be brittle, difficult to work with, and expensive.

A. Gallium Arsenide

Gallium arsenide-based cells have been under development since the mid-1950s[18] to exploit their potential for high efficiency, low performance degradation with increasing temperature, and tolerance to ionizing radiation total dose. These attributes are related to gallium arsenide's relatively large direct-type bandgap: 1.42 eV vs 1.12 eV for Si. They make GaAs cells good candidates for concentrator and space-based environments. Equation (2.1-1) and the values for the effective densities of states (Table 1) indicate GaAs has a smaller intrinsic concentration n_i than silicon at any temperature. The consequent low saturation current density J_0 produces large values for V_{oc} which lead to high efficiency. For the common situation where recombination is dominated by base recombination, GaAs cells achieve J_0 values as small as 4 x 10^{-20} A/cm^2 at

COMPARISON OF HIGHEST MEASURED 1-SUN CELL EFFICIENCIES FOR Si, GaAs, AND MONOLITHIC TWO-TERMINAL TANDEM JUNCTION CELLS
(25 °C and Global AM1.5 Spectrum)

		sc-Si	mc-Si	GaAs Bulk	GaAs/Ge	GaInP/GaAs Tandem
Area	(cm^2)	4.0[1]	1.0[1]	3.9[2]	4.0[2]	0.25
V_{oc}	(mV)	709	628	1022	1035	2385
J_{sc}	(mA/cm^2)	40.9	36.2	28.2	27.6	14.0
FF	(%)	82.7	78.5	87.1	85.3	88.5
η	(%)	24.0	17.8	25.1	24.3	29.5

1 *Aperture area for a submodule*   2 *Total cell area*

Table 7   Comparison of highest measured 1-sun cell efficiencies for Si, GaAs, and monolithic two-terminal tandem junction cells. (25°C and global AM1.5 spectrum)

300 K – much smaller than the 2.5 to 5 x 10^{-14} A/cm^2 seen in the best Si cells[19,20,21]. Low J_0 values also produce relatively high V_{OC} values at higher temperatures than are practical for Si cells. Short lifetimes seen in the base of the best GaAs cells (tens of nanoseconds vs hundreds of microseconds for Si) and the inability of GaAs to absorb sunlight below 1.42 eV (longer than λ = 0.9 μm) lead to low values for J_{SC}. This keeps resistive losses low, and the fill factor tends to be high. Table 7 compares record efficiencies for single-crystal Si, multicrystalline Si, and GaAs-based cells[22].

During crystal growth, an over-pressure of arsenic is required to maintain stoichiometry of the solid. Arsenic vapor is in equilibrium with the melt and solid for an arsenic vapor pressure of about 1 atm. If an over-pressure of arsenic is not maintained, the liquid arsenic condenses, and the melt becomes gallium rich. In the 1970s, bulk GaAs was often produced by the horizontal Bridgeman technique in which a GaAs single-crystal seed is sealed in a quartz ampoule with a GaAs charge at one end of the ampoule and a supply of arsenic at the other end[23]. The ampoule is moved through a two-temperature zone rf-heated furnace. The GaAs charge is held in the high temperature zone (about 1240°C) so that it remains in a molten state. The arsenic supply is held in the cooler zone (about 630°C) to maintain an arsenic vapor pressure of about 1 atm. The seed is in between the two zones and contacts the melt. As the ampoule slowly moves toward the low-temperature end of the furnace, a single-crystal ingot develops from the seed[24]. Five-kilogram ingots have been grown. This technique produces low-dislocation density ingots, but the diameter is limited to several centimeters.

Bulk GaAs wafers are now more commonly produced by the liquid-encapsulated Czochralski (LEC) technique[25] in which a liquid B_2O_3 cap floats on top of the GaAs melt in a Czochralski puller. The B_2O_3 cap flows at about 450°C and seals the Ga and As charge before the As starts to sublimate. Synthesis of GaAs starts at about 800°C with the reaction $Ga_{liquid} + As_{solid} = GaAs_{solid}$. The GaAs melts at 1238°C and is held just above this temperature to maintain stoichiometry near that of the compound. Growth is initiated by dipping a seed through the B_2O_3 cap and into the melt. The B_2O_3 cap and high argon overpressure prevent depletion of arsenic in the melt through sublimation and subsequent loss of stoichiometry in the crystal. High purity LEC wafers several inches in diameter are routinely produced[26,27,28]. LEC GaAs is grown in both quartz and pyrolytic boron nitride crucibles, the latter of which can be reused several times. Growth is about five to ten times as expensive as the growth of Czochralski silicon wafers of the same size.

There are quite a few variations on the Bridgeman and Czochralski techniques. However, because of the complicated growth environment and the inherent scarcity of gallium, the price of GaAs wafers will probably always be considerably greater than the price of Si wafers. As of 1992, electronic-grade Ga costs about $425/kg with a world-wide production of 8 metric tons per year[29]. This is compared with electronic-grade Si scrap feedstock, which is roughly $10/kg and is produced in the thousands of metric tons.

The problems of feedstock price and brittleness are ameliorated by growth of very thin single-crystal GaAs films on foreign substrates – most notably Ge. Germanium has approximately the same density as GaAs (5.32 g/cm³), but is less brittle. Substrates as thin as 100 μm are self-supporting and can be handled like conventional Si cells. The large absorption coefficient of GaAs makes thin films economically attractive[30]. Close lattice match between Ge (a = 0.5658 nm) and GaAs (a = 0.5653 nm) allows single-crystal growth on the Ge substrate. This has an immediate cost advantage as Ge substrates are only about 30% as expensive as GaAs substrates. Growth of high quality GaAs films on Si substrates is desirable but has not been accomplished. GaAs films have been epitaxially grown on Si, but the 4% lattice mismatch between GaAs and Si makes the material rich in defects and the method is cumbersome[31]. Also, GaAs on Si requires the (less preferred) p⁺n polarity as Si is a donor impurity. Germanium is an amphoteric dopant in GaAs.

Bulk GaAs cells and GaAs/Ge cells with passive substrates have approximately equal radiation hardness[32]. The short absorption length and short lifetime decouple the performance of the cell from crystal damage produced by high energy particles more than a few microns into the surface. Low energy particles can still damage the cell close to the junction. A 50-μm cover glass is sufficient to shield the cell from 2 MeV protons. Referenced to their starting efficiencies, GaAs/Ge cells with passive substrate interfaces are more resistant to electrons and protons than are Si cells. EOL-to-BOL power ratios for 1-MeV fluences of 10^{14} cm⁻² electrons are about 1.5 times that of Si cells[33]. An active interface between a GaAs thin film and a Ge substrate increases the rate of degradation compared to a passive interface where there is no rectifying junction to separate carriers. Degradation severity increases with the width of the cell. Thick cells have more volume per unit top-area in which to accumulate particle damage which serves as a sink for minority carriers[34].

Early GaAs cells were made by diffusion of impurities into bulk wafers. Silicon and zinc have distribution coefficients greater than 0.1 and are common n- and p-type dopants, respectively. Selenium (n-type) and beryllium (p-type) are also common dopants. Performance for these early cells was poor, with V_{oc} about 0.55 V and efficiency less than 7%. This was traceable to several problems. Substrates had high defect densities that produced low minority carrier lifetimes. The combination of high surface recombination velocity at the unpassivated front surface and the high absorption coefficient resulted in large emitter contributions to the dark current. High series resistance from unreliable ohmic contacts to n-type material degraded short-circuit current. Low shunt resistance from unintentional impurities near the junction degraded open-circuit voltage. Fill factor was degraded by high R_s and low R_{sh}.

Modern GaAs-based cells have reduced these problems by using high purity substrates and low specific contact resistance metallizations. Metallizations to n-type GaAs include sputtered Au/Ge/Ni. Nickel enhances the diffusion of Ge in GaAs and allows Ge to fill Ga lattice sites near the surface[35]. This produces low interfacial contact resistance, $\sim 10^{-6}$ Ω cm^2. Thick Au plating on top yields low sheet resistance for the contact fingers. However, the biggest improvement in GaAs cell performance has come with the addition of a transparent thin (~ 0.03 μm) heavily-doped single-crystal p-type $Al_xGa_{1-x}As$ *window layer* that passivates the front surface of a p-GaAs/n-GaAs cell[36]. While efficiencies of 20% have been achieved with shallow homojunction GaAs cells without windows, the addition of an $Al_xGa_{1-x}As$ window layer raises the efficiency to as much as 25%. A heavily-doped p$^+$ GaAs cap is used directly under the metallization to provide good ohmic contact to the $Al_xGa_{1-x}As$. Front contact metallization is Au (95%)/Zn (5%)[37]. In between the metallization fingers, the cap is etched away with H_2O_2/NH_4OH or citric acid solution. Low mole fractions that do not require a GaAs cap are also feasible[38]. The generic $Al_xGa_{1-x}As$-window cell[39] incorporates an n-type back-surface field layer. Ohmic contact to the n-type GaAs base is made with Au(84%)/Ge(12%)/Ni(4%). Contacts are annealed at 400°C for 1 min. An evaporated 0.09-μm layer of Sb_2O_3 or PECVD 0.078-μm layer of SiN$_x$ provides an anti-reflection coating. A wider deeper reflection minimum for $Al_xGa_{1-x}As$ is achieved with two layers, MgF_2/ZnS or MgF_2/SiN_x[40].

Though the $Al_xGa_{1-x}As$ layer forms a heterojunction with the GaAs underneath, its primary function is that of a surface passivant rather than a collecting junction. The p-type GaAs layer serves as the cell's emitter. Aluminum mole fraction x is 0.8 to 0.9. Lattice mismatch of $Al_xGa_{1-x}As$ to GaAs is much less than 1% for all values of x. This makes monolithic growth of $Al_xGa_{1-x}As$ easy. The $Al_xGa_{1-x}As$ bandgap type, though, is a function of the Al mole fraction. As Al content increases, the bandgap undergoes a change from the direct bandgap of GaAs to the indirect bandgap of AlAs. The transition occurs at approximately x = 0.4. With increasing x, the magnitude of the bandgap monotically increases from 1.42 eV (pure GaAs) to 2.16 eV (pure AlAs). As an example, for x = 0.8, $Al_xGa_{1-x}As$ is an indirect bandgap material with E_g = 2.1 eV. This makes it transparent to a large part of the solar spectrum that can be absorbed by GaAs.

For zero crystal electron momentum (fig. 2.1-6), the valence and conduction bands are 2.6 eV apart. In this sense, $Al_{0.8}Ga_{0.2}As$ can be thought of as having a direct bandgap of 2.6 eV as well as an indirect bandgap of 2.1 eV. Photons between 2.1 eV and 2.6 eV will be weakly attenuated as in other indirect bandgap materials. Above 2.6 eV, the photons raise valence band electrons directly to the conduction band and the absorption rate is large. Consequently, to maximize the transparency of the $Al_xGa_{1-x}As$ window across the entire GaAs-usable part of the solar spectrum, it is necessary to make the window layer very thin – 0.01 to 0.05 μm. Additionally, a very thin window allows some of the minority electrons generated in the $Al_xGa_{1-x}As$ to diffuse to the $Al_xGa_{1-x}As$/ GaAs interface before recombining, thus enhancing the high-energy spectral response. The window introduces little strain to the GaAs surface while providing a transparent conductive film that lowers both the series resistance and the surface recombination velocity at the emitter surface.

Growth techniques for GaAs cells with a window layer include liquid-phase epitaxy (LPE)[41] and metalorganic chemical vapor deposition (MOCVD)[42,43]. Both methods allow growth of sequential single-crystal (epitaxial) thin films on a similar substrate or on a lattice-matched foreign substrate such as Ge. Film thicknesses are as small as several nanometers. MOCVD is the preferred method as it allows improved control of dopant concentration and the option for gradual changes in the film constituents. Gradual changes minimize stress in the film. The thinner layers that can be grown by MOCVD in a single process step result in higher efficiencies. A third possible method for epitaxial growth is molecular-beam epitaxy (MBE) in which constituent elements are evaporated under extremely low pressures (10^{-10} Torr) and travel with few collisions directly to a lattice-matched substrate surface. MBE is not a serious candidate for cell fabrication because of the expense and slow throughput of ultra-high vacuum equipment.

Liquid-phase epitaxy creates an epitaxial layer by deposition from a molten solution which is saturated at the growth interface. An example is the growth of a GaAs epilayer from a Ga-rich solution which is saturated with As. Temperatures are around 800°C. Growth is based on the property that the solubility of a dilute constituent in a liquid solvent decreases with decreasing temperature. It begins when the saturated solution is brought into contact with the relatively cool substrate. Growth rate is limited by the rate at which constituents can diffuse to the interface between the liquid and the evolving epilayer. Advantages of LPE include low dislocation density (lower than the substrate), high reproducibility, and the convenience of only a few growth parameters. These are the cooling rate, growth time, and the substrate orientation, composition, and temperature. LPE is a fast inexpensive growth technique, but shows poor control because of fast dopant diffusion at high temperature. There are several techniques for starting and terminating growth. The most common is the horizontal sliding technique. In this approach, the substrate is held in a shallow depression in a flat graphite holder. The growth solutions are held in horizontally arranged graphite chambers that sit on top of the substrate holder and are free to move across the holder. Each chamber has a hole in the bottom. As a chamber moves across the graphite holder, the growth solution is brought into contact with the substrate.

In MOCVD, sometimes called organometallic vapor phase epitaxy or OMVPE, a metal organic of one element and the hydride of the other element are mixed in the vapor phase. Pyrolysis on the heated substrate yields a thin film of the desired semiconductor compound. The general equation for the most common class of MOCVD reactions is

$$R_nM + XH_n \rightarrow MX + nRH$$

where R is an organic radical, M and X are the components of the resultant semiconductor, and n is an integer. Examples are

$$(CH_3)_3Ga + AsH_3 \rightarrow GaAs + 3\ CH_4$$

and

$$TMGa + TMAl + AsH_3 \rightarrow Al_xGa_{1-x}As + CH_4$$

where TMGa is trimethyl gallium and TMAl is trimethylaluminum. Reaction temperatures are in the range of 600° and 750°C. Growth rate for the GaAs reaction is 1-10 µm/hr. This is a fast growth rate considering the need for only thin layers. The MOCVD growth apparatus uses an rf coil to heat the substrate without significantly heating the walls of the growth chamber. This cold-wall feature of MOCVD results in little residual dopant in the chamber. The dopant atoms that have not been incorporated into the growth are expelled when the chamber is vented, allowing precise doping transitions not available in LPE. Carbon, silicon, and zinc are the primary residual unintentional impurities.

The ease of producing complicated profiles with MOCVD makes the method well suited to bandgap-engineered high efficiency devices. An example is the growth of ternary compound profiles such as $Al_{0.8}Ga_{0.2}As/Ga_{0.83}In_{0.17}As_{0.75}P_{0.25}$ on GaAs or Ge substrates. The AM1.5 efficiency for this tandem structure is over 18%[44]. Another example is bandgap grading in the growth of a $Al_{0.15}Ga_{0.85}As/Si$ tandem-junction device with

AM0 efficiency of about 20%[45]. For this device, a p+nn+ AlGaAs top cell is grown on a p+nn+ Si bottom cell with a GaAs buffer separating the two cells. The emitter of the AlGaAs upper cell has a graded bandgap in which the Al mole fraction decreases, in sequential layers, from x = 0.30 at the surface to x = 0.15 at the emitter-base junction. These successive layers are between 20 and 50 nm thick. By grading the Al mole fraction, a built-in electric field is created in the bulk of the emitter that improves the excess photogenerated carrier collection of the top cell. Complicated reproducible MOCVD growths are common. MOCVD has also been used to produce simple structures on potentially inexpensive and large-area substrates such as an $Al_xGa_{1-x}As$/GaAs cell on polycrystalline Ge[46]. Efficiency is about 16% AM1.5, but 20% for these substrates is anticipated.

Perhaps the most significant achievement of MOCVD in photovoltaics is the development of the world-record *monolithic two-terminal* tandem-junction cell[47]. This impressive $Ga_{0.5}In_{0.5}P$/GaAs structure (referred to simply as GaInP/GaAs), discussed in section 4.6, achieves AM1.5 efficiencies of 29.5% under one sun, and over 30% under 115-260 suns. Substrates are either bulk GaAs or Ge. The top cell is n+p GaInP with an n-type AlInP surface passivating window layer and a bandgap of 1.85 eV. The bottom cell is n+p GaAs with an n-type GaInP window and a bandgap of 1.42 eV. Total thickness of the active layers is about 5 μm. The top cell absorbs the blue part of the spectrum, while the bottom cell absorbs some of the light which passes through the top cell and which is more toward the red part of the spectrum. The cells are interfaced by very heavily-doped layers that serve as tunnel junctions and allow carriers to pass from one cell to the other like an ohmic contact. Tandem cells are very challenging structures because of the numerous design trade-offs between the two subcells. These include the choice of polarity. Less efficient GaInP/GaAs tandem cells with p+n polarity for each subcell have been investigated[48]. Regardless of polarity, the bandgap for the lower subcell is fixed at 1.42 eV. But the ternary nature of the upper subcell allows adjustment of its bandgap. Various growth parameters affect the ordering of Ga atoms in the GaInP lattice, and ordering is a determinator of energy gap magnitude. For GaInP, the bandgap can be adjusted from 1.82 eV to 1.89 eV by varying the growth temperature from 650°C to 600°C[49]. Precise tailoring of the upper subcell bandgap affords maximum absorption for the combination of the two subcells.

Because the subcells are in series, their open-circuit voltages are additive. However, J_{sc} for the overall cell is fixed to the lesser of the two subcell short-circuit currents. The thickness of the upper subcell can be adjusted so that its J_{sc} equals that of the lower subcell. This is the condition for maximum overall J_{sc} because if the upper subcell has a larger J_{sc} than that of the lower one, the excess current capacity is wasted. The optimum upper bandgap and thickness combination is spectrum-dependent. Through the upper subcell's J_{SC}, the overall J_{SC} is related to the upper subcell's E_g, V_{OC}, and FF. Large E_g favors V_{OC} and FF; small E_g favors J_{SC}. Individual layer thicknesses in the MgF_2/ZnS anti-reflective coating are chosen so that the resultant reflection minimum is at the wavelength for maximum J_{SC}. Thus, the tandem cell's performance is dependent on a careful adjustment of many different design parameters. It is characterized by J_{SC} of about half that for a conventional GaAs cell, and a V_{OC} more than twice that of a GaAs cell. Low J_{SC} prevents large resistive losses in the emitter and the metal contacts, and this helps to keep the fill factor large. Figure 5.2-1 shows the AM0 tailored profile for the tandem cell[50].

Tandem cells in the various material systems do not have to be monolithic. They can be mechanically *stacked*, in which the upper and lower cells are partly or totally independent of each other. Record efficiency for all solar cells is held by a mechanically stacked GaAs/GaSb structure. It achieves 32.6% efficiency under 100 suns AM1.5. As with monolithic tandems, the upper cell absorbs at the

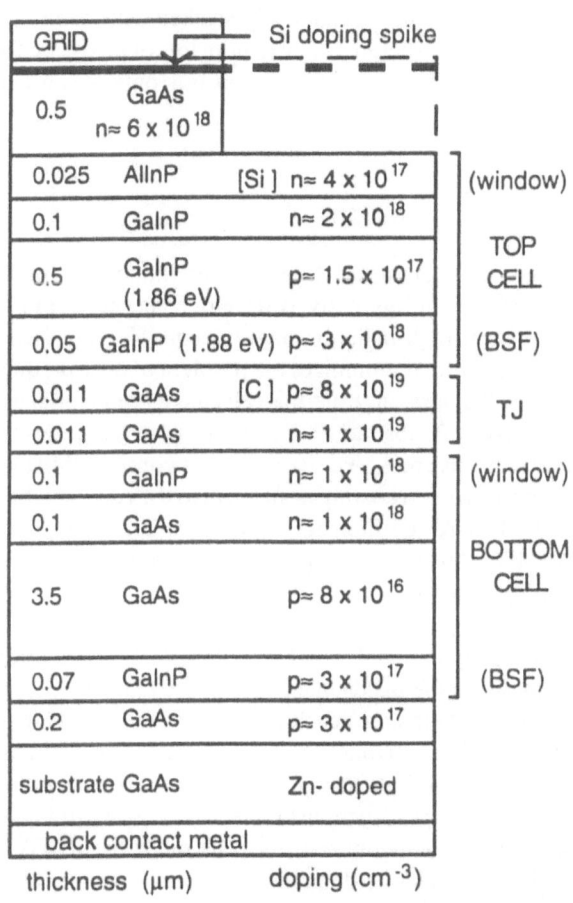

Fig. 5.2-1   Profile of $Ga_{0.5}In_{0.5}P$/GaAs tandem junction cell tailored for AM0 spectrum. (*after ref. 50*)

blue end of the spectrum, while the lower cell absorbs the red light that passes through the upper cell. Mechanically stacked cells can have two, three, or four terminals. (Monolithic cells can have three terminals, but this is rare.) For the four-terminal mechanical configuration, there is great freedom to optimize the design of the individual cells. The three-terminal configuration has a wire connecting the base of the upper cell with the emitter of the lower cell which serves as a third terminal. The two-terminal configuration connects the upper base with the lower emitter, and has the current matching constraint of a tandem monolithic structure. While mechanically stacked cells avoid the problems of growing complicated epitaxial structures, they require lots of hardware for positioning and connecting the cells. Consequently, this approach is only feasible in concentrator systems.

B. Indium Phosphide

During the mid-1990s, GaAs-based cells started to eclipse Si cells for space deployment. However, crystalline InP cells are being aggressively developed because of their almost flat radiation degradation curves in high-radiation environments[51,52]. This makes their EOL efficiency higher than GaAs/Ge and minimizes the problem of heat-production from excess power at the start of the mission. Bulk InP cells have been fabricated by diffusion of sulfur (In_2S_3) into p-type InP substrates[53]. The p-type base is doped with Zn to 2-8 x 10^{16} cm^{-3}. Front ohmic contacts are Ag/Pd. Back contact consists of plated Ag on top of Ag/Zn. A SiN_x/ZnS anti-reflection coating is matched to the cell's spectral response. Cells of this type with 16-17% AM0 efficiency were flown in 1990 in lunar orbit on the Japanese Muses-A satellite. After 60 days in orbit, there was essentially no degradation as indicated by a flat V_{OC} curve.

InP is a heavy (4.8 g/cm^3) direct bandgap material with E_g = 1.35 eV. The absorption coefficient is > 10^4/cm for photons above the bandgap energy. Emitters must be very shallow – on the order of 50 nm – for photons to reach the base. This requirement, coupled with the expense of bulk InP (at least 30 times Si), makes epitaxy on Si, GaAs, Ge, or InP substrate a better choice than bulk diffusion. The expense of bulk InP is partly attributed to the need for high-pressure Czochralski pullers necessary to maintain stoichiometry during ingot growth. Homoepitaxial InP cells made by MOCVD achieve AM0 efficiencies over 19%. Efficiency degradation is less than 5% after 10^{15} cm^{-2} 1-MeV electrons and less than 10% after 10^{12} cm^{-2} 10-MeV protons. Heteroepitaxial growth on Si substrates is inherently less expensive because of the relatively low capital investment associated with MOCVD equipment. Heteroepitaxial InP cells on Si substrates are more radiation resistant than the homoepitaxial variety and have the additional advantages of lighter weight and greater strength. Typical be-ginning-of-life (BOL) efficiency is low – only about 10%. However, the efficiency decays to only 9% absolute after 10^{16} cm^{-2} 1-MeV electrons[54]. Beyond this fluence, these cells have higher efficiencies than GaAs/Ge cells with 22% initial efficiency. The lower BOL efficiency is attributed to the 8% lattice mismatch between Si and InP. Minority carrier lifetime is limited by the high dislocation density. By EOL, the life-time is radiation-damage limited, and InP/Si becomes superior to GaAs/Ge.

To achieve the approximately 19% BOL efficiencies seen in homoepitaxial cells, while maintaining the radiation hardness of the InP/Si cell, the strain from the large lattice mismatch must be relieved. An approach well-suited to MOCVD is the growth of a GaAs buffer followed by a series of $In_{1-x}Ga_xAs$ layers (strained-layer superlattice)[55] on the Si substrate. This is not to be confused with a tandem structure. The mole fraction is slowly varied from 1 at the GaAs interface to a value of 0.47 at the InP interface. At $x = 0.47$, $In_{1-x}Ga_xAs$ is latticed matched to InP. A major problem is the diffusion of Si as a donor impurity into the GaAs buffer during the 1000°C process. The p-on-n structure at the GaAs/Si interface is a reverse diode for holes travelling toward the back contact. This is relieved by creating a tunnel diode at the interface. An $InP/In_{1-x}Ga_xAs/GaAs/Si$ structure using a superlattice to reduce strain and a Si substrate to lower cost is a lead-ing candidate for the next generation space cells. BOL efficiencies of 14% have been achieved.

Several laboratories have made small $InP/In_{0.53}Ga_{0.47}As$ monolithic tandem cells with pre-irradiated efficiencies exceeding 20% 1-sun AMO. The best of these was *over 22%*[56]. In early experiments[57], efficiency degraded quickly after 10^{14} cm^{-2} 1-MeV electrons. By the time the fluence reaches 10^{15} cm^{-2} 1-MeV electrons, the efficiency of the tandem $InP/In_{0.53}Ga_{0.47}As$ cell approaches that of the best heteroepitaxial n$^+$p InP/Si cell – about 12.5%. The severe efficiency degradation is attributed to unequal J_{sc} degradation rates in the two subcells, with the lower subcell degrading faster and becoming dominant. Closer rates of current degradation can be obtained by adjusting the base doping in the lower subcell. This will allow high performance tandem cells with essentially zero degradation up to at least 3×10^{14} cm^{-2} 1-MeV electrons. For maximum EOL perfor-mance, the design goal in any tandem structure requires the subcells to be current matched at EOL.

Radiation hardness in InP is attributed to two distinct self-annealing mechanisms, ther-mal annealing and minority carrier injection annealing, that repair lattice displacement damage. Unlike GaAs, which requires temperatures above 150°C, thermal annealing in InP occurs at room temperature. Thermal energy in the InP lattice is sufficient to allow displaced atoms to return to their original sites. Minority carrier injection annealing occurs while the cell is forward-biased from illumination. Self-annealing is a complicated mechanism involving several defect levels[58] and the doping impurity. With self-annealing, short-circuit current of the cell recovers to almost its pre-irradiation value. When deployed in space, n⁺p InP cells experience both thermal annealing and illumination induced annealing. That is, the radiation resistance is improved while the cells are illuminated. Similar processes occur in GaAs cells, but GaAs has higher migration energies for radiation-induced defects than does InP, and the rate of self annealing is much slower.

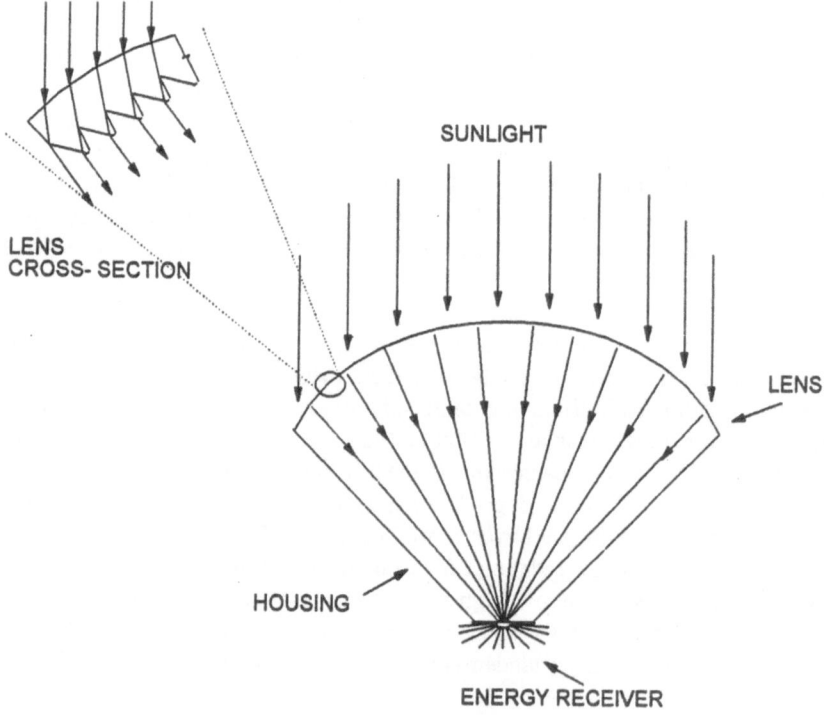

Fig. 5.3-1   Cross section of 21X linear concentrator module. a) Schematic showing curved Fresnel lens and aluminum heat sink under the solar cells.

5.3 CONCENTRATOR SYSTEMS

Concentrator systems trade expensive semiconductor for cheap plastic by using a Fresnel lens to focus incoming sunlight into either a linear or a disk shape that covers the solar cell. Originally, concentrator systems used convex lenses to focus light to a disk shape, but this quickly proved to be a poor approach because of the weight, expensiveness, and low concentration-ratio limitation of the lenses. Parabolic troughs have been used for linear focus systems. Since the 1970s, most systems have been designed around Fresnel lenses (either flat or curved) in which the light is focused by a refraction grating (fig. 5.3-1). Unlike convex lenses that refract light through a large mass of mate-

b) Segment from a 12-ft module. (*Photos courtesy ENTECH, Inc., Keller, Texas*)

rial, Fresnel lenses have a thin cross section and are much less expensive. For high-concentration ratio systems (> 100 X), semiconductor area is very small and it is expected to be cost effective to use high-efficiency Si or GaAs-based cells with optimized evaporated-metal grids to handle high current densities. Systems with low concentration ratios (~ 20 X) can use more conventional screen-printed Si cells that are only slightly modified by way of the grid metallization. A duplex arrangement of 21X modules that use conventional Si cells is shown in fig. (5.3-2). Each module in fig. (5.3-2) has the cross section shown in fig. (5.3-1). For 19% efficient Si cells, the module efficiency and output are 15.8% and 480 W (25°C AM1.5 spectrum), respectively. Calculations for a 100-MW array with this technology indicate a system/field operational ac-power efficiency of 12.1% (at 60°C) and a deployed area of 3.6 km².

Fig. 5.3-2   Duplex arrangement for 21X concentrator modules. (*Photo courtesy ENTECH, Inc., Keller, Texas*)

Cell efficiency is enhanced by any degree of concentrated illumination. This is attributed to the logarithmic increase in V_{oc} as J_{sc} increases with the concentration factor. However, for n⁺p silicon cells, experiment shows that V_{oc} does not increase as quickly with J_{sc} as predicted by eq. (3.5-5)[59]. The explanation for this is complicated and involves an effective saturation current density for the high-injection regime. Overall, the device physics of cells under concentration are not well-understood. This is reflected in the observation that some heterojunction devices with mediocre 1-sun efficiencies may show dramatic improvement under low concentration (~ 20X). Thin-film InP cells on GaAs substrates have shown efficiencies (at AM0 spectrum, 25°C) change from 13.7% at 1-sun to 18.7% at 40 suns[60]. For any cell, the increase in V_{oc} with concentration is partly offset by higher temperature if the cell is not equipped with an effective heat sink. Under sufficiently high concentration, recombination is dominated by Auger processes. Auger recombination is a three-body process whereby an energetic electron makes a band-to-band transition to recombine with a hole. The difference in energy is transferred to another electron or to a hole. This is contrasted with radiative recombination common in III-V materials where a photon is emitted, and with recombination through intermediate states (Shockley-Read-Hall recombination) that predominates in Si in low injection. When Auger recombination is dominant, the diode curve factor, n, becomes less than one. Analysis and empirical data[61] suggest that V_{oc} for a cell under concentrated light is maximized for a base resistivity of approximately 0.3 Ω cm. In fact, most silicon cells intended for high-concentration have low base resistivities (0.15 to 0.2 Ω cm). For concentration factors in the 20 to 300 range, silicon cell efficiency improves about 3 to 5 percentage points absolute for constant temperature.

Offsetting these advantages are three distinct problems. First, concentrator systems require *solar tracking*. In general, the performance of any photovoltaic system is enhanced by tracking. In the limiting case of a 1X concentrator, i.e., a flat-plate system, a static elevation or tilt of the module equal to the latitude approximates one-axis (declination) tracking. This provides maximum average annual direct insolation without actual tracking. For example, if the location is 23° from the equator, then by fixing the tilt of the module to 23° above the horizon, the average annual direct insolation is maximized to the extent possible without tracking. The performance, of course, could be further improved by tracking along the azimuth. For flat-plate systems, performance displays a cosine dependence on the tracking angle along each axis. However, for concentrator systems, precise two-dimensional tracking is imperative for high-ratio systems, and common in low-ratio systems. Off-axis tracking of one or two degrees drastically lowers the output of a concentrator system. Sometimes this is ameliorated with a refractive glass secondary optical element that redirects the light onto the cell in the event of mis-pointing.

Tracking can be accomplished by passive devices, for example, a closed-fluid mechanism in which an unevenly heated fluid expands. As the fluid redistributes itself, the center of mass of the module-tracker assembly changes and the module is tilted more towards the sun. Active tracking is more effective and affords the precise orientation required by high-ratio systems. An active tracking approach common in the 1970s was a real-time sensor-based system that measured light intensities from different directions. This provided a differential signal that controlled a motor. By the 1990s, sensor-based systems were replaced by more accurate microprocessor-controlled tracking in which reference is made to stored data on exact solar position. Since tracking is a process of slow-adjustment, only small motors are necessary for simultaneously steering an array of modules. Two-dimensional tracking in which position is periodically adjusted by two small stepping motors is successfully used in a commercial 100-kW array of concentrator modules[62] (fig. 5.3-3). The system is microprocessor controlled and relies on stored data of the sun's position throughout the year. Regardless of the tracking mechanism used by any manufacturer, tracking requires mechanical components and this tends to lower long-term system reliability compared to flat-plate systems.

Fig. 5.3-3   100-kW array of 21X concentrator modules at Ft. Davis, TX.
(*Photo courtesy ENTECH, Inc., Keller, Texas*)

An issue related to tracking is the large reflective loss caused by grid metallization. To minimize series resistance from large currents, concentrator cells may need over 35%

top-side metallization. Incoming light is reflected off the metal lines and is wasted. This problem is solved with prismatic plastic covers that are oriented parallel to the metal lines and positioned over each cell. Light is directed away from the metal lines and onto the active area of the cell, greatly enhancing performance (fig. 5.3-4).

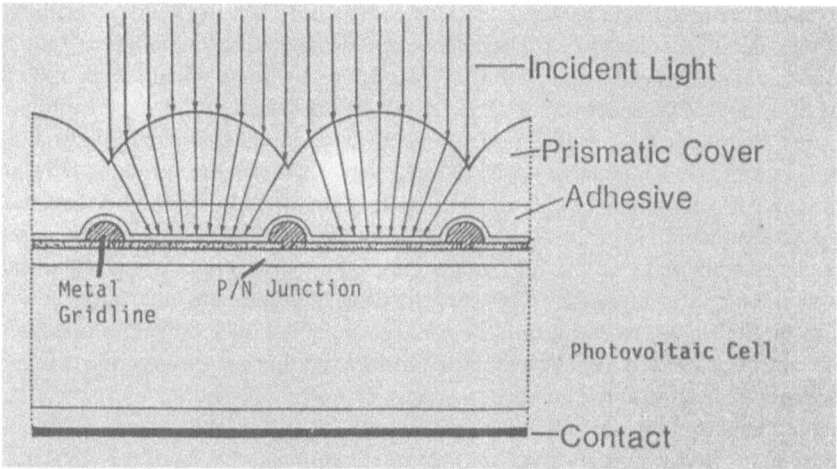

Fig. 5.3-4   Cross section of prism cover showing refraction of light away from metal grid lines. (*Photo courtesy ENTECH, Inc., Keller, Texas*)

Secondly, concentrators require a cooling mechanism to avoid high cell temperatures and subsequent severely degraded open-circuit voltages and efficiencies. Due to its narrower bandgap, Si suffers more in this regard than does GaAs. Thus, high concentration cells tend to be GaAs-based. However, even GaAs cells must be kept well under 100°C for reasonable efficiencies. Cells can be either passively or actively cooled, with the passive heat sink option being the most popular. This takes the form of large-area cooling fins thermally interfaced to the back of the cell. The equipment associated with tracking and cooling adds expense to the system. Both of these problems are intrinsic to the system in that they can be dealt with through innovative engineering. The third problem, though, is extrinsic to the system and represents the biggest obstacle to concentrators. This is the geographically-based requirement of high *direct normal* (orthogonal) insolation. Concentrators display very poor performance in scattered or diffuse lighting environments. They are restricted to sunny locations with few clouds. An ideal location is the southwestern U.S. where the direct normal insolation exceeds 2600 kWh/m²/yr in certain sections, e.g., most of Arizona[63]. This is compared with much of the eastern U.S., where the direct normal insolation is less than 1500 kWh/m²/yr, and less than 1200 kWh/m²/yr throughout New England.

In spite of these problems, concentrators have respectable performance in suitable locations. Under 100-300 suns, commercial *systems* made on silicon float-zone material are expected to have field operating efficiencies around 17%[64]. In 1995, a pre-commercial 100 X module achieved 21.6% efficiency at 45°C operating temperature[65]. The cell for this module achieved a concentration record for a commercial silicon cell: 26.8% under 50 suns AM1.5 spectrum at 25°C. By 1996, GaAs cells achieved 27.6% under 255 suns, and monolithic GaInP/GaAs tandem junction cells reached 30.2% at 180 suns. The tandem cell maintained over 29% efficiency at high concentration (425 suns). Such high efficiency under high concentration is particularly significant because it suggests the potential for very cost-effective systems. It is anticipated that the tandem junction cell technology now being commercialized for space applications will eventually be employed in terrestrial high-concentration modules. An interesting possible application for high-concentration PV systems is solar energy storage through hydrogen production. In such a system, intense infrared light passes through the cell (or is split from the spectrum) to superheat a supply of water. The shorter wavelengths are converted to current by the cell and supply energy for electrolysis, which is very efficient at temperatures on the order of 1000 K. Electromagnetic energy is changed to chemical energy.

Aside from increased top-side metal coverage, series resistance in Si cells is controlled through thick grid lines ($\rho_s < 0.01$ Ω /□). These are formed by electroplating Ag on top of evaporated metal, i.e., several microns Ag on top of 30 nm Ti/50 nm Pd/100 nm Ag. As with all high efficiency cells, evaporated TiPdAg is preferred. The Ti film adheres strongly to the highly-doped silicon surface and forms a silicide with low interfacial contact resistance. A thick layer of silver allows low sheet resistance for the line as a whole. The palladium serves as a diffusion barrier to prevent silver from spiking through the titanium film and possibly shunting the junction. Numerical modelling is used to determine optimal design relationships between grid spacing, linewidth, and performance for a particular cell[66]. Series resistance can be further lowered by increasing the emitter thickness to over 1 μm, but this degrades spectral response and J_{sc}.

Maintenance of J_{sc} is particularly significant because of data suggesting *superlinearity* with illumination intensity at concentrations over 100 suns for *some* n⁺p Si cells with low base resistivity[7]. This is attributed to conductivity modulation in the base resulting from large carrier generation during concentrated illumination. The effect is dependent on the concentration ratio. For medium ratios, the current is not so great that overall series resistive loss dominates the increased conductivity in the base, and J_{sc} remains slightly superlinear with the intensity. For high ratios (several hundred X), the current density is very large and the overall series resistive loss more than offsets the enhanced base conductivity. In that case, J_{sc} becomes sublinear with the intensity. These nonlinear effects have been observed in GaAs cells[67] and in GaInP/GaAs tandem cells[68], as well as in Si cells, but they are not well-understood. The fill factor reflects the series resistance and increases with intensity until resistive losses become large. The open-circuit voltage is related to the intensity through J_{sc}, and rises in a roughly logarithmic fashion with intensity.

Accurate determination of efficiency at high concentrations is made problematic by the difficulty in calibrating the light source. One-sun intensity is easily calibrated according to the short-circuit current from a standard reference cell. However, this approach is not possible for concentrated light because the light spot becomes nonuniform across the area of the cell. Inaccurate calculations will result if the reference and test cells have different topology. Even if the cells have similar topology, there is still a question about the calibration of the reference cell's J_{sc} with intensity. The problem of calibration of light intensity has usually been dealt with by measuring the test cell's J_{sc} for one sun and then assuming that it scales linearly with intensity. As seen above, this assumption is unwarranted and leads to underestimates for many concentrator efficiencies. An accurate method has been developed for calibrating the light source which uses the measured spectral response of the cell under different intensities[67]. Such measurements indicate that concentration-dependent variations in the back surface recombination velocity and diffusion length are contributing factors to the nonlinearity between J_{sc} and intensity in Si cells. Nonlinearity is shown to be specific to various regions of the spectral response curve. Similar factors are shown to control the nonlinear J_{sc} in GaAs cells.

5.4 THERMOPHOTOVOLTAIC DEVICES

Not all photovoltaic cells are intended for operation under sunlight. In the emerging technology of *thermophotovoltaics* (TPV), cells are exposed to infrared light from an artificial source. The source, referred to as an emitter (*not to be confused* with the emitter of the cell) is usually an oxide ceramic, or silicon carbide, mantle. One side of the emitter faces the TPV cells. The reverse side of the emitter is heated by burning a hydrocarbon fuel. Radioisotopes can also be used to heat the emitter for spacecraft-based TPV systems. In either case, the semiconductor must be spectrally matched to the heat source for reasonable heat-to-electric efficiency. If a large portion of the spectrum has photon energy slightly greater than the bandgap of the absorbing material, the conversion efficiency can be large.

The first TPV system was developed in 1963 with germanium cells and a propane fueled emitter[69]. Early TPV designs were envisioned as large systems using concentrated sunlight to heat an emitter to the temperature required for good spectral match and high emissive power[70,71]. For these solar TPV designs, minimum efficiency for competitiveness with other PV systems was estimated to be very high, as much as 40%. Interest in TPV waned until the early 1990s when several companies started to investigate the possibility of small TPV systems that could be cost competitive at more reasonable efficiencies. Solar TPV has been abandoned. Recent research focuses on hydrocarbon-fueled emitters. This redirection is driven both by new technologies and new market applications. The new technologies include semiconductors capable of absorbing long-wavelength radiation and ceramic emitters with desired emission spectra. Possible applications include electric vehicles and quiet power sources for remote military equipment. TPV is also envisioned as a co-generation technology, providing both heat and electric power in remote locations. Renewed interest in TPV is reflected by the increase in research papers seen in the mid-1990s[72].

Central concepts in all studies of thermal radiation are that of a *blackbody* and blackbody radiation[73]. A blackbody is defined as a source which displays perfect, i.e., the maximum possible, emission and absorption at all wavelengths. In thermal equilibrium, the emission and absorption of the blackbody at any given wavelength are equal. The irradiated or *emissive power*, sometimes referred to as exitance, has the units of power density per unit bandwidth. It is given by Planck's spectral distribution relation, and is a function of temperature only for a given wavelength. While the concept of a blackbody is an ideality, it can be approximated by a cavity having a small orifice and held at a constant temperature. Electromagnetic radiation entering the cavity undergoes numerous reflections from the interior walls and is almost entirely absorbed. Thus, such a construction is almost a perfect absorber. Since it is held in thermal equilibrium, absorption equals emission, and the cavity is almost a perfect emitter. Various opaque materials roughly approximate a blackbody. For these materials, radiation is concentrated in the infrared for temperatures between about 10 and 5000 K. The emissive power of a real body compared to that of a blackbody is described by the body's *emissivity*. The emissivity or emittance is defined as the ratio of the power density per unit bandwidth emitted into a hemisphere by the body in question to that emitted by a blackbody at the same temperature. Thus, a blackbody has emissivity equal to one at all wavelengths, while a real body always has emissivity less than one at any particular wavelength. By the Stefan-Boltzmann law, the emissive power integrated over all wavelengths, or total emissive power, increases as T^4, where T is the Kelvin temperature. The surface of the sun approximates a blackbody at roughly 6000 K (fig 1.3-1). While the total emissive power of the sun at its surface[74] is 7360 W/cm^2, the vast distance to the earth (and absorption in the earth's atmosphere) reduces this value to about 0.1 W/ cm^2 at the earth's surface. In comparison, cells close to an artificial 2000 K blackbody source see a total emissive power of about 90.9 W/cm^2. Thus, in terms of total power density seen by the cells, TPV systems are intrinsically high concentration systems.

Blackbody spectral composition varies only with temperature. As the temperature decreases, the peak of the blackbody spectrum, i.e., the wavelength for maximum power density per unit bandwidth, shifts toward longer wavelengths (fig. 5.4-1). This means that, as the temperature of the blackbody decreases, less and less of the total emissive power of the source will be in the form of photons with energy exceeding the bandgap of the semiconductor. Photons with energy below the bandgap are not absorbed, while photons far above the bandgap energy are wasted through thermalization as electron kinetic energy is changed into lattice vibrations. These energetic photons contribute little to the cell's efficiency because they can create

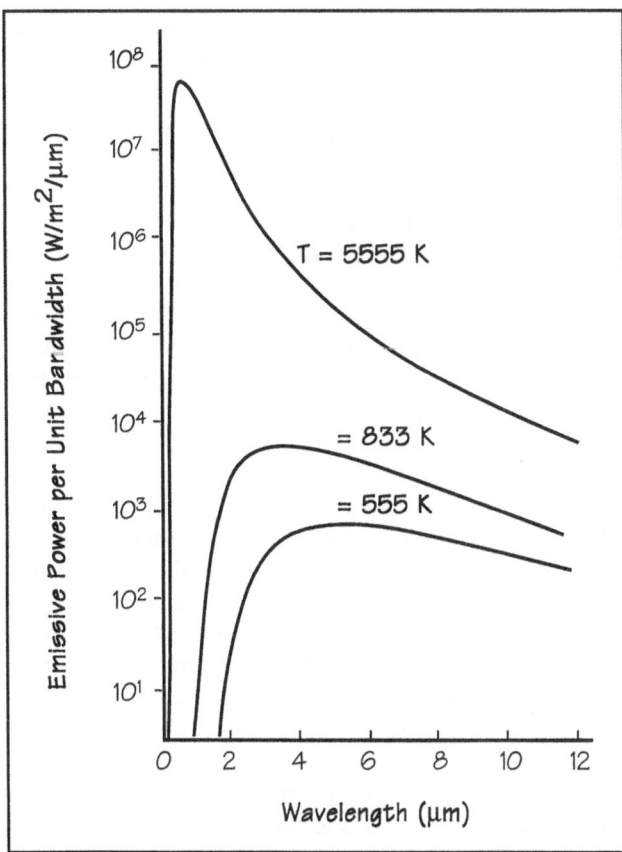

Fig. 5.4-1   Blackbody emissive power for several temperatures. (*adapted from ref. 73*)

only one electron-hole pair. As an example, for silicon, both 0.3 μm- and 0.9 μm-wavelength photons create an electron-hole pair. But in terms of the photon-to-electric energy ratio, the 0.9-μm photon contributes three times as much to the efficiency of the cell. Additionally, because of the inherent concentration in a TPV system, energetic photons tend to quickly heat the semiconductor and lower cell efficiency. As with other photovoltaic systems, ideal radiation for thermophotovoltaic conversion (assuming sufficient thickness for absorption) is at an energy just slightly above the bandgap. For Si cells at 300 K, this is 1.12 eV, or 1.11 μm. These spectral considerations have an *interesting consequence* for TPV development. While lower-temperature blackbody sources have a smaller fraction of usable (convertible) power density than higher-temperature sources, they also have relatively less wasted power in that fraction which is convertible. As an example, silicon cells facing a 6000 K blackbody radiation source (filtered to AM1.5 and one-sun) have a limit conversion efficiency of about 26%, versus about 56% for a 2000 K source (fig. 5.4-2). This suggests that blackbody sources at temperatures much lower than that of the sun (1300-2200 K) can drive highly efficient TPV systems if the unconvertible low-energy photons are reflected back to the emitter for reabsorption[75]. Filters with dielectric coatings that reflect wavelengths longer than those corresponding to the bandgap, or quartz windows which are highly reflective at wavelengths longer than about 2.5 μm, are commonly used.

The reabsorption of these photons by the emitter helps maintain high emitter operating temperature, and subsequent high emissive power. Heat-to-electric efficiency of 40% is plausible for a 2000 K blackbody emitter facing high-efficiency silicon cells, under the condition that 95% of the sub-bandgap photons are reflected. Long-wavelength photons can also be *recuperated* by passing the fuel/air line through the burner exhaust. By doing this, some of the unconvertible long-wavelength energy is transferred back to the emitter. Recuperation can also be integrated into an active cooling system for the back surfaces of the cells, so that cell temperature is kept acceptably low. This is particularly important for narrow-bandgap cells, since the intrinsic concentration and J_0 go up quickly with temperature, and V_{oc} drops considerably. Because of poor reabsorption and recuperation of unconvertible photons, and other engineering problems, the several commercial (pre-commercial) systems that have been built have heat-to-electric efficiencies well below the limit values.

Artificial radiators with temperatures comparable to the surface of the sun are unfeasible and, from the above considerations, not necessarily desirable. Furthermore, there are no true blackbody sources. Thermophotovoltaic systems attempt to achieve useful efficiencies with much lower temperatures and non-blackbody sources. There are two generic approaches.

In the first approach, a narrow bandgap material is matched to a relatively low-temperature (1100° to 1550°C) *graybody* emitter[76,77]. A graybody is defined as an object which has a uniform, though less than unity, emissivity over all of the spectrum. In practice, graybodies have an approximately constant (but less than unity) emissivity over a wide portion of the spectrum. Silicon carbide emitters are good examples of graybodies, with emissivity approximately 0.8. By "matched," it is meant that the semiconductor and emitter are chosen so that the semiconductor's bandgap is close to the

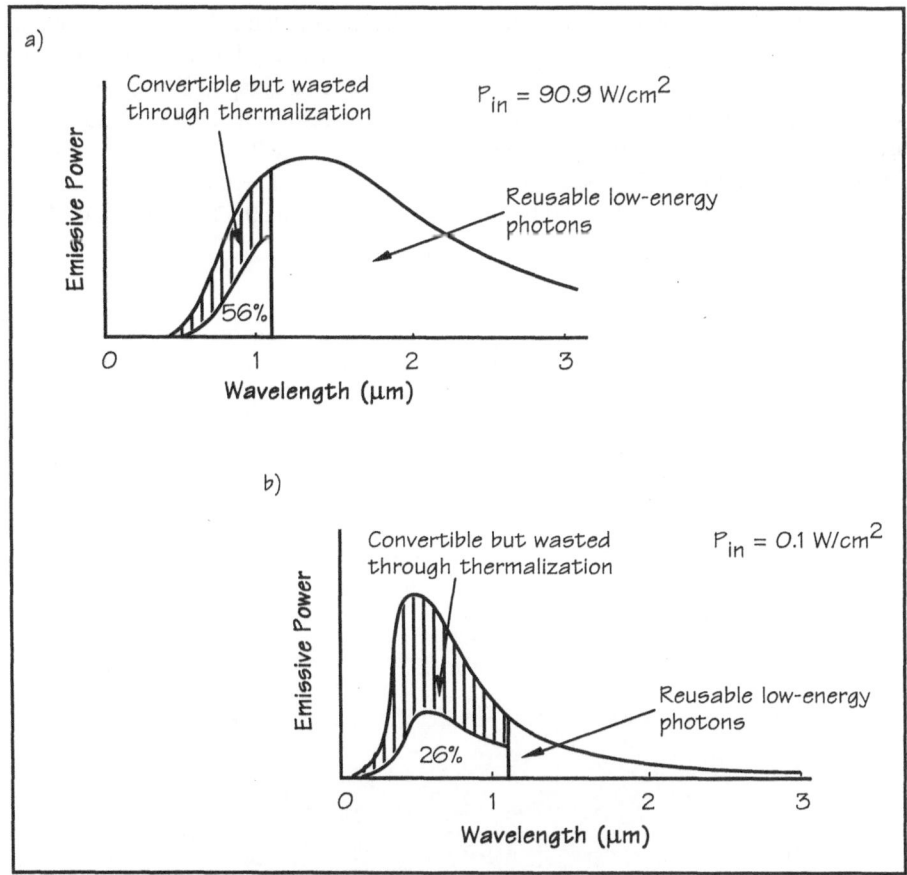

Fig. 5.4-2   Comparison of the convertible, but wasted, fractions of the 2000 K and 6000 K blackbody power density spectra interfaced to silicon TPV. a) 2000 K close proximity source, b) 6000 K sunlight at earth's surface. (*adapted from ref. 75*)

peak of the emitter spectrum for the operating temperature. For example, silicon would not be a good choice for a graybody operating at 2000 K because the peak is at 1.5 μm, which is well beyond the silicon bandgap of 1.12 eV (λ = 1.1 μm). For the 2000 K graybody, GaSb would be a better match because its bandgap is 0.72 eV (λ = 1.7 μm). Graybody spectra from oxide ceramics result primarily from free-carrier heat absorption in the emitting material. Suitable emitter materials have the properties of high melting point, oxidation stability, and thermomechanical strength. They include Si_3N_4, SiC, alumina (Al_2O_3), and stabilized zirconia (ZrO_2)[78]. Silicon carbide is frequently used, but is limited to about 1500°C, above which rapid oxidation becomes a problem. Aside from GaSb, semiconductor candidates for the relatively low-temperature long-wavelength graybody approach include $In_xGa_{1-x}As$. The bandgap of $In_xGa_{1-x}As$ can be adjusted by varying the mole fraction x and allows a careful match to the emitter spectrum. For growth on InP substrates, acceptable values of x are 0.78 to 0.53, corresponding to 0.5 eV and 0.74 eV, respectively[79]. Reabsorption and recuperation make a big difference. As an example, V_{OC} and FF for GaSb cells fall from 0.52 V and 0.84 to about

Fig. 5.4-3   100-W GaSb-based propane-fueled TPV system. (*Photo courtesy JX Crystals, Inc.*)

0.49 V and 0.72, respectively, as the cell temperature climbs from 25°C to about 70°C. A successful GaSb-based TPV system which uses heat recuperation by passing the combustion air through the burner exhaust is now commercialized (fig. 5.4-3). IR filters are attached to the individual cells to reject long wavelengths. The cells are exposed to a 1400°C SiC graybody emitter so that they produce over 1 W/cm². System output is about 100 W.

In the second approach, the semiconductor is matched to a high-temperature (1400°C to 1800°C) *selective emitter* that has a strong emission peak corresponding to an energy just above the bandgap. Selective emitters have non-uniform emissivities. These

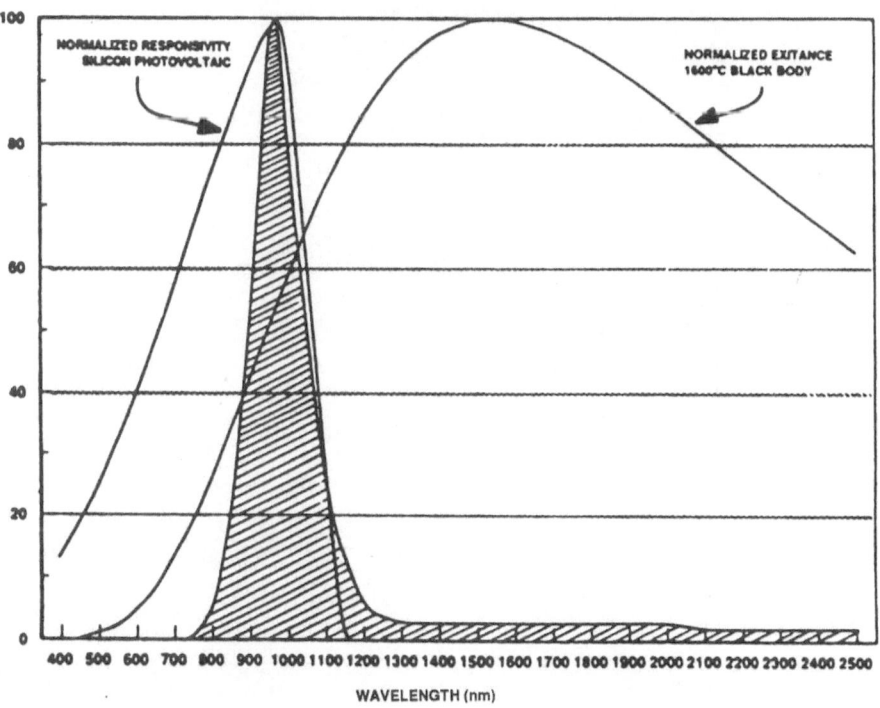

Fig. 5.4-4   Normalized spectral exitance of fibrous ytterbia emitter (cross-hatched) compared to Si cell responsivity and 1600°C blackbody spectrum (*Photo courtesy Thermolyte Corporation*)

Fig. 5.4-5
20-W silicon-
based ytterbia
emitter TPV
system. a) Sys-
tem showing fuel
tank, *left*.
b) Close-up of
burner in
operation, *below*.
(*Photos courtesy
Thermolyte
Corporation*)

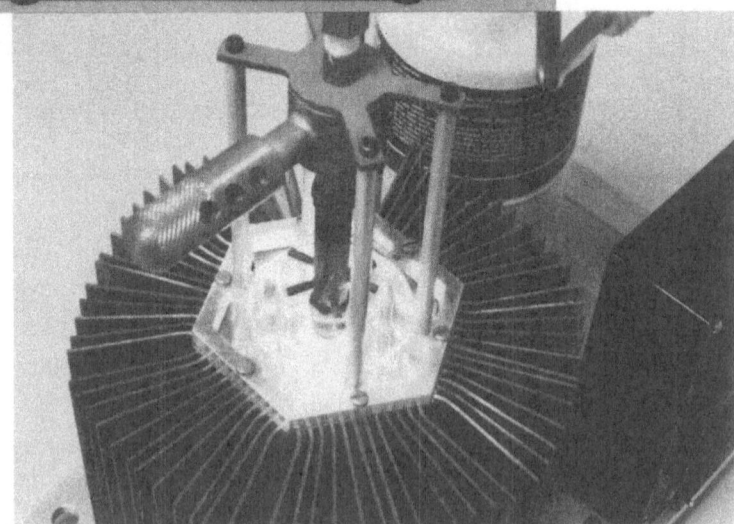

emitters are usually oxide ceramics of rare-earth elements. Like most oxide ceramics, their spectra contain broad emissions in the UV and far-IR that can be described as graybody spectra. However, characteristic only of the rare-earth elements, these oxides also have several strong narrow emission peaks in the visible and near-IR. These peaks result from electron energy transitions limited to vacancies in the inner electron shells. For very thin sections of these ceramics (~ 10 μm), emission of the graybody part of the spectrum is *suppressed*, allowing only the narrow emission peaks in the visible and near-IR to appear. As a result, by using the appropriate rare earths and by controlling geometry, mantles can be made so that their spectra are characterized by the presence of only a few strong emission peaks in the near-IR[80]. Oxides of interest include neodymia (with an emission peak at 2.4 μm), holmia (2.0 μm), erbia (1.55 μm), and ytterbia (0.98 μm). Erbia and ytterbia are particularly interesting for TPV[81]. Their spectra each have one strong emission peak. An erbia emitter is well-matched to cells made from germanium ($E_g = 0.66$ eV which corresponds to a wavelength of 1.88 μm). Ytterbia (Yb_2O_3) is well-matched to silicon[82]. The normalized spectral exitance of a fibrous ytterbia emitter is superimposed on the normalized responsivity of a high-efficiency silicon cell in fig. 5.4-4. The figure includes the normalized exitance of a 1600°C blackbody to illustrate the large fraction of unconvertible power in that spectrum. The ytterbia peak is narrow, with a full-width-at-half-maximum of 0.15 μm. For selective emitters, the wavelength position of the peak is independent of the temperature. However, because of the narrowness of the peak, high temperatures (1700°C or higher) and large emitting surface areas are necessary to achieve radiated power densities of at least several watts per square centimeter. For the peaks of the ceramic oxides, the power density per unit bandwidth goes up much faster than T^4.

Rare-earth oxide ceramics are stable in a flame environment and they can be shaped into forms with large effective surface area, for example, ceramic fiber bundles. Special techniques are necessary to prevent failure of the ceramic structure from thermal stress or poor mechanical strength[83]. One procedure, in particular, has been developed which produces high-strength ytterbia fiber bundles anchored to a planar porous substrate of cordierite (Celcor)[84]. A successful system based on this technology and using high-efficiency silicon cells is shown in fig. 5.4-5. The fiber bundle structure starts with a rayon textile precursor. A continuous yarn, consisting of 200-500 rayon filaments, is impregnated with an aqueous nitrate salt of ytterbium. The yarn is threaded through the pores of the substrate so that small loops extend several millimeters above the surface. The yarn is then dried and subjected to a high temperature bake which converts the metal salt into the oxide, ytterbia, while removing the rayon through pyrolysis. The result is a skeleton of 10-μm diameter ceramic filaments arranged in loops (bundles). The several hundred loops per square inch have a very high total effective

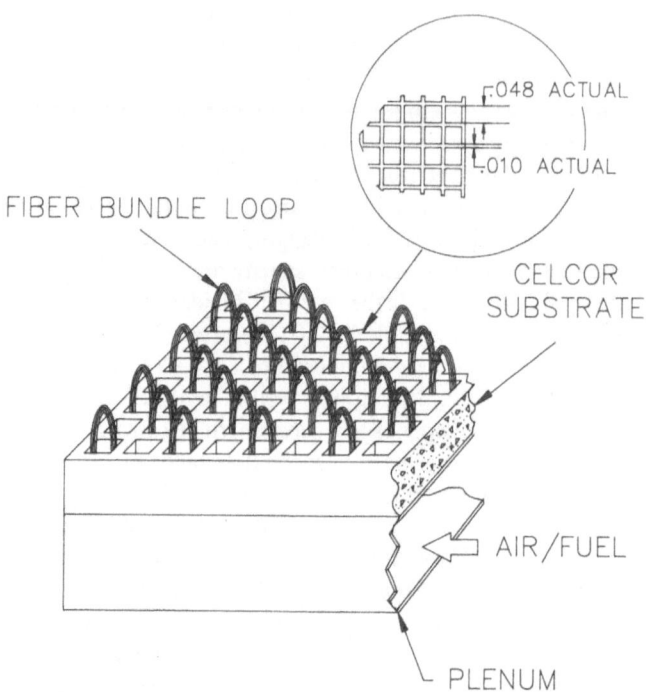

Fig. 5.4-6   Ytterbia fiber bundle emitter structure. (*Photo courtesy*
Thermolyte Corporation)

surface area (fig. 5.4-6). There is little thermal stress in the filaments during temperature cycling due to the small diameter. When heated by a propane gas flame which
extends through the pores and just slightly above the substrate surface, the fiber bundles
produces a silicon-convertible power density of up to 6 W/cm^2. This represents a *usable*-illumination concentration of considerably more than 60 suns. Short-circuit current densities reach ~ 10 A/cm^2.

An alternative to either the graybody or selective emitter approaches, announced in
1996, is an IR emitter that is matched to the entire response band of GaSb. Its power
density spectrum is shown in fig. 5.4-7 and compared to a blackbody emitter. This
proprietary matched IR emitter yields improved GaSb system efficiency compared to a
SiC graybody emitter.

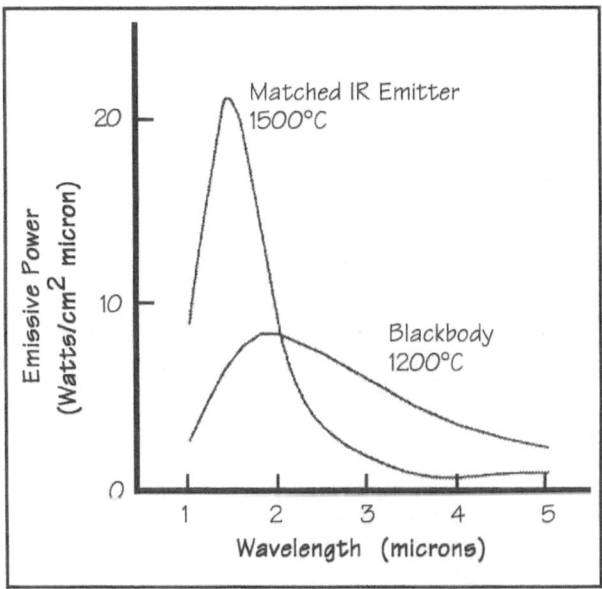

Fig. 5.4-7   Power spectrum of an IR emitter very well matched to the GaSb response band. (*Courtesy JX Crystals, Inc.*)

There are several trade-offs between the graybody and selective emitter approaches. Graybodies can have high convertible emissive power at relatively low temperatures. Selective emitters can achieve high emissive power only through high temperature. This requirement becomes stronger as the emission peak becomes narrower. Low temperatures are easier to work with because there is less likelihood of thermal stress in the emitter structure, ytterbia fiber bundles notwithstanding. However, compared to selective emitters, graybodies tend to waste a lot of power in long wavelengths and the need for reabsorption and recuperation becomes very important. Additionally, the narrowness of selective emitter spectra makes those emitters ideally suited to various semiconductors. This includes re*latively* low-cost high-efficiency silicon cells. For systems with silicon cells matched to ytterbia operating at 1810°C, the thermal-to-electric efficiency has been measured[85] at 41%. Graybody emitters are not suited to silicon because the temperature would have to be extremely high for reasonably efficient operation. For both graybody and selective emitter systems, overall fuel-to-electric efficiency is about 5% to 10%. Because of their high thermal-to-electric efficiency capability, selective emitter systems *might* have a greater engineering potential for overall fuel-to-electric efficiency. But this is not clear as of 1996. It is probably the case that both approaches will find many commercial applications. A logical application is co-generation of heat and electric power for remote residential and military field operations (fig. 5.4-8). Another application possibility is low-NO_x-emission electric vehicles. At the temperatures associated with graybody emitters, nitrogen oxide production from the burning of propane is low. Above 1600-1700°C, NO_x concentrations are expected

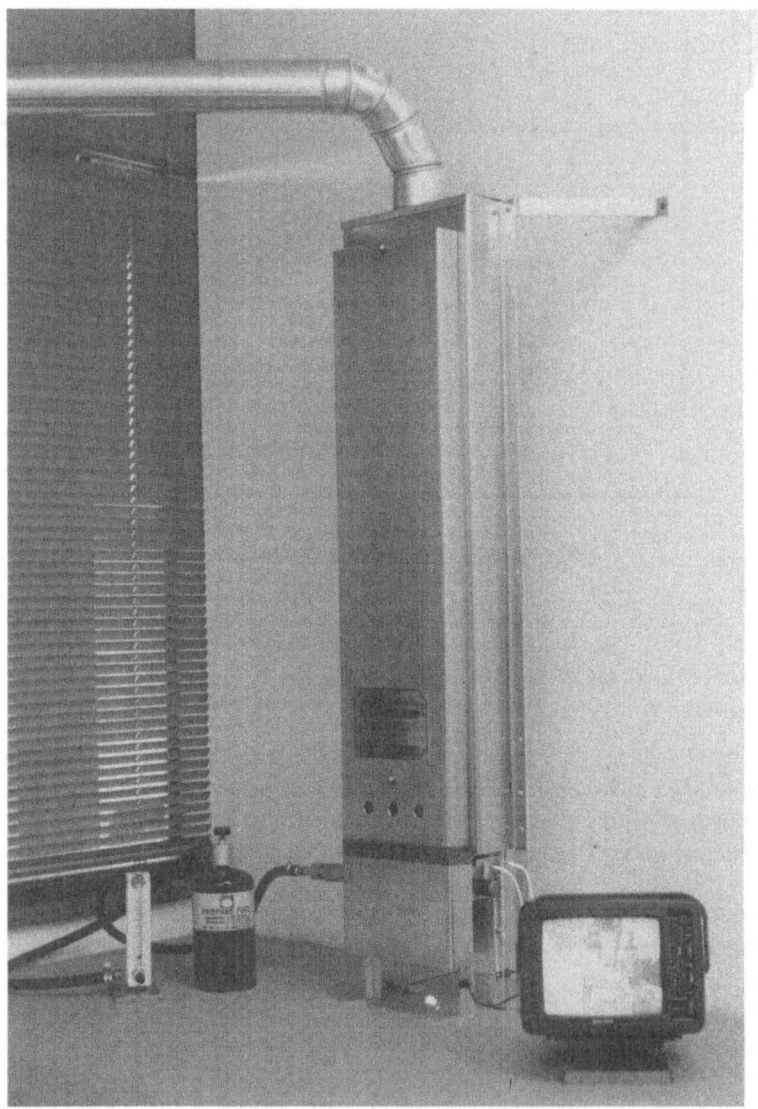

Fig. 5.4-8   GaSb-based 30-W (electric), 440-W (thermal), TPV
residential co-generation unit. (*Photo courtesy JX Crystals, Inc.*)

to rise exponentially with temperature and reach several hundred parts per million. Measured results indicate that this does not happen with ytterbia fibers. Ytterbia appears to suppress the NO_x formation process, perhaps through catalysis that lowers the combustion temperature of the blue part of the flame. Concentrations remain on the order of only several ppm. Thus, both graybody and selective emitter approaches may prove useful for low-emission electric vehicles or other applications requiring low-pollution low-noise electric generation. Like many developments in photovoltaics, TPV has an exciting future.

REFERENCES

[1]   *Photovoltaic News*, ed. by P.D. Maycock, vol. 15, p. 4, Feb. 1996.

[2]   D.E. Carlson, U.S. Patent No. 4,064,521, "Semiconductor device having a body of amorphous silicon," Dec. 20, 1977.

[3]   K. Zweibel, H. Ullal, and B. von Roedern, "Progress and issues in polycrystalline thin-film PV technologies," preprint of paper for 25th IEEE Photovoltaic Specialists Conf., May 1996.

[4]   Two special issues on the future of thin-film solar cells in *Progress in Photovoltaics*, vol. 3, Sept-Oct,1995, and Nov-Dec, 1995.

[5]   D.L. Staebler and C.R. Wronski, "Reversible conductivity changes in discharge-produced amorphous Si,", *Appl. Phys. Lett.*, vol. 31, p. 292-294, Aug. 1977.

[6]   A. Catalano, "Solar cells made of amorphous and microcrystalline semiconductors," in *Amorphous & Microcrystalline Semiconductor Devices*, ed. by J. Kanicki, vol. 1, Chap. 2, Artech House, Boston, 1991.

[7]   R.W. Sanderson, et al., "Performance of silicon solar cells in the concentration range of 150-1500 suns," *Conf. Rec. of the 17th IEEE Photovoltaic Specialists Conf.*, pp. 1309-1313, May 1984.

[8]   T.F. Ciszek, et al, "Growth of thin crystalline silicon layers for photovoltiac device use," *J. Crystal Growth*, vol. 128, pp. 314-318, 1993.

[9]   A.M. Barnett, "Thin polycrystalline silicon solar cells on low cost substrates," *Conf. Rec. 6th Intl. Photovoltaic Science and Engineering Conf.*, pp. 737-744, 1992.

[10]   T.F. Ciszek, "Techniques for the crystal growth of silicon ingots and ribbons," *J. Crystal Growth*, vol. 66, pp. 655-672, 1984.

[11]   S.A. Edmiston, et al., "Modelling of thin-film crystalline silicon parallel multijunction solar cells," *Progress in Photovoltaics*, vol. 3, pp. 333-350, Sept-Oct, 1995.

[12] A. Eyer, et al., "Crystal structure and electrical properties of silicon sheets grown from powder (SSP-method)", *Conf. Rec. of the 19th IEEE Photovoltiac Specialist Conf.*, pp. 951-954, May 1987.

[13] T.F. Ciszek, et al., "Si thin layer growth from metal solutions on single-crystal and cast metallurgical-grade multicrystalline substrates," *Conf. Rec. of the 23rd IEEE Photovoltaic Specialists Conf.*, pp. 65-72, May 1993.

[14] N. Sridhar, et al., "Polysilicon films of high photoresponse, obtained by vacuum annealing of aluminum capped hydrogenated amorphous silicon," *J. Appl. Phys.*, vol. 78, pp. 7304-7311, Dec 1995.

[15] M.J. Kardauskas, et al., "The coming of age of a new PV wafer technology - some aspects of EFG polycrystalline silicon sheet manufacture," preprint of paper for 25th IEEE Photovoltaic Specialist Conf., May 1996.

[16] T.F. Ciszek, "The growth of silicon ribbons for photovoltaics by edge-supported pulling (ESP)," in *Silicon Processing for Photovoltaics I*, ed. by C.P. Khattak and K.V. Ravi, Elsevier Science Publishers, New York, 1985.

[17] T.F. Ciszek, "Silicon for solar cells," in *Crystal Growth of Electronic Materials*, ed. by E. Kaldis, Elsevier Science Publishers, New York, 1985.

[18] D.A Jenny, J.J. Loferski, and P. Rappaport, "Photovoltaic effects in gallium arsenide pn-junctions and solar energy conversion," *Phys. Rev.*, vol. 101, pp. 1208-1209, Feb. 1956.

[19] G.B. Lush, et al., "Determination of minority carrier lifetimes in n-type GaAs and their implications for solar cells," *Conf. Rec. of the 22nd IEEE Photovoltaic Specialists Conf.*, pp.182-187, Oct. 1991.

[20] S.R. Kurtz, J.M. Olson, and A. Kibbler, " High efficiency GaAs solar cells using GaInP$_2$ window layers," *Conf. Rec. of the 21st IEEE Photovoltaic Specialists Conf.*, pp. 138-140, May 1990.

[21] J. Zhao, et al., "717 mV open-circuit voltage silicon solar cells, " *Proc. of the 11th E.C. Photovoltaic Solar Energy Conf.*, pp. 272-275, Oct. 1992.

[22] M.A. Green, et al., "Solar cell efficiency tables (version 7)," *Progress in Photovoltaics*, vol. 4, pp. 59-62, Jan-Feb, 1996.

[23] A.R. Von Neida and J.R. Nielsen, "Synthesis and crystal growth of GaAs and GaP for substrates," *Solid State Technology*, vol. 17, pp. 90-98, Apr. 1974.

[24] T.R. AuCoin and R.O. Savage, "Bulk growth of gallium arsenide," in *Gallium Arsenide Technology*, Chap. 2, Howard W. Sams & Co., Indianapolis, 1985.

[25] J.B. Mullin, B.W. Stranghan, and W.S. Brickell, "Liquid encapsulation techniques: the use of an inert liquid in suppressing dissociation during the melt-growth of InAs and GaAs crystals," *J. of Phys. and*

Chem. of Solids, vol. 26, pp. 782-784, Apr. 1965.

[26] M.E. Weiner, D.T. Lasota, and B. Schwartz, "Liquid encapsulated Czochralski growth of GaAs," *J. Electrochem. Soc.*, vol. 118, pp. 301-306 , Feb. 1971.

[27] T.R. AuCoin, et al, "Liquid encapsulated compounding and Czochralski growth of semi-insulating gallium arsenide," *Solid State Technology*, vol. 22, pp. 59-67, Jan. 1979.

[28] *Semiconductors and Semimetals, vol. 20*, ed. by R.K. Willardson and A.C. Beer, pp. 163-167, Academic Press, New York, 1984.

[29] *Mineral Commodity Summaries 1993*, U.S. Department of the Interior, Bureau of Mines, pp. 64-65, 1993.

[30] M. Braun, et al., "GaAs solar cells: structure and technology," *Conf. Rec. 22nd IEEE Photovoltaic Spec. Conf.*, pp. 377-380, Oct. 1991.

[31] Y. Okada, et al., "Growth of GaAs and AlGaAs on Si substrates by atomic hydrogen-assisted MBE (H-MBE) for solar cell applications," *Conf. Rec. of the First World Conf. on Photovoltaic Energy Conversion*, pp. 1701-1704, Dec. 1994.

[32] B. Bollani, et al., Development and optimization of GaAs/Ge single junction solar cells," *Conf. Rec of the 22nd IEEE Photovoltaic Specialists Conf.*, pp. 295-298, Oct. 1991.

[33] G.C. Datum and S.A. Billets, "Gallium arsenide solar cells -a mature technology," *Conf. Rec. of the 22nd IEEE Photovoltaic Specialists Conf.*, pp. 1422-1428, Oct. 1991.

[34] D.J. Flood and I. Weinberg, "Advanced solar cells for satellite power systems," *Optoelectronics*, vol. 9, pp. 451-458, Dec. 1994.

[35] M.N. Yoder, "Ohmic contacts in GaAs," *Solid State Electronics*, pp. 117-119, 1980.

[36] H.J. Hovel and J.M. Woodall, "Theoretical and experimental evaluations of $Ga_{1-x}Al_xAs$-GaAs solar cells," *Conf. Rec. 10th IEEE Photovoltaic Spec. Conf.*, pp. 25-30, Nov. 1973.

[37] C. Amano, et al., "Fabrication and numerical analysis of AlGaAs/GaAs tandem solar cells with tunnel interconnections," *IEEE Trans. on Electron Devices*, vol. 36, pp. 1026-1035, June 1989.

[38] M.E. Klausmeier-Brown, "Low-bandgap front surface barriers for GaAs solar cells," *Conf. Rec. of the 22nd IEEE Photovoltaic Specialists Conf.*, pp. 220-222, Oct. 1991.

[39] K.A. Bertness, et al., "16% efficient GaAs solar cell after 10^{15} cm^{-2}, 1-MeV radiation," *Conf. Rec. of the 21st IEEE Photovoltaic Specialists Conf.*, pp. 1231-1234, May 1990.

[40] A. Baldus, et al., "GaAs one-sun and concentrator solar cells based on LPE-ER grown structures," *Conf. Rec. of the First World Conf. on Photovoltaic Energy Conversion*, pp. 1697-1700, Dec. 1994.

[41]   J.J. Hsieh in *Handbook on Semiconductors*, ed. by S.P. Keller, vol. 3, Chap. 6, North-Holland Publishing Co., New York, 1980.

[42]   H.M. Manasevit, "Recollections and reflections of MO-CVD," *J. Crystal Growth*, vol. 55, pp. 1-9, 1981.

[43]   P.D. Dapkus, et al., "High purity GaAs prepared from trimethylgallium and arsine," *J. Crystal Growth*, vol. 55, pp. 10-23, 1981.

[44]   P.R. Sharps, et al., "GaInAsP lattice matched to GaAs for solar cell applications," *Conf. Rec. of the 22nd IEEE Photovoltaic Specialists Conf.*, pp. 315-317, Oct. 1991.

[45]   M. Umeno, et al., "High efficiency AlGaAs/Si tandem solar cell over 20%," *Conf. Rec of the First World Conf. on Photovoltaic Energy Conversion*, pp. 1679-1684, Dec. 1994.

[46]   R. Venkatasubramanian, et al., "Material and device characterization toward high-efficiency GaAs solar cells on optical-grade polycrystalline Ge substrates," *Conf. Rec. of the First World Conf. on Photovoltaic Energy Conversion*, pp. 1692-1696, Dec. 1994.

[47]   D.J. Friedman, et al., "30.2% efficient GaInP/GaAs monolithic two-terminal tandem concentrator cell," *Progress in Photovoltaics*, vol. 3, pp. 47-50, Jan-Feb, 1995.

[48]   P.R. Sharps, et al., "Development of p-on-n GaInP$_2$/GaAs tandem cells," *Conf. Rec. of the First World Conf. on Photovoltaic Energy Conversion*, pp. 1725-1728, Dec. 1994.

[49]   S. Kurtz and D. Friedman, tutorial notes on III-V fundamentals and applications, 25th IEEE Photovoltaic Specialists Conf., May 1996.

[50]   K.A. Bertness, et al., "High efficiency GaInP/GaAs tandem solar cells for space and terrestrial applications," *Conf. Rec. of the First World Photovoltaic Energy Conf.*, pp. 1671-1678, Dec. 1994.

[51]   I. Weinberg, "Radiation damage in InP solar cells," *Solar Cells*, vol. 31, pp. 331-348, Sept. 1991.

[52]   I. Weinberg, et al., Effects of radiation on InP cells epitaxially grown on Si and GaAs substrates," *Conf. Rec of the 21st IEEE Photovoltaic Specialists Conf.*, pp. 1235-1240, May 1990.

[53]   M. Yamaguchi, et al., "First space flight of InP solar cells," *Conf. Rec. of the 21st IEEE Photovoltaic Specialists Conf.*, pp. 1198-1202, May 1990.

[54]   S.J. Wojtczuk, "Development of InP solar cells on inexpensive Si wafers," *Conf. Rec. of the First World Conf. on Photovoltaic Energy Conversion*, pp. 1705-1708, Dec. 1994.

[55]   C.J. Keavney, et al., "Fabrication of n+/p InP solar cells on silicon substrates," *Appl. Phys. Lett.*, vol. 54, pp. 1139-1141, 1989.

[56]   M.W. Wanless, et al., "Improved large-area two-terminal InP/Ga$_{0.47}$In$_{0.53}$As tandem solar cells," *Conf. Rec. of the First World Conf. on Photovoltaic Energy Conversion*, pp. 1717-1720, Dec. 1994.

[57]   R.J. Walters, et al., "1-MeV electron irradiation of monolithic two-terminal InP/Ga$_{0.47}$In$_{0.53}$As solar cells," *Conf. Rec. of the 23rd IEEE Photovoltaic Specialists Conf.*, pp. 1475-1478, May 1993.

[58]   A. Sibile, "Origin of the main deep electron trap in electron irradiated InP," *Appl. Phys. Lett.*, vol. 48 pp. 593-595, Mar 1986.

[59]   J.G. Fossum and F.A. Lindholm, "The dependence of open-circuit voltage on illumination level in pn junction solar cells," *IEEE Trans. on Electron Devices*, vol. ED-24, pp. 325-329, Apr 1977.

[60]   M.W. Wanlass, et al., "High-efficiency heteroepitaxial InP solar cells," *NASA Conf. Pub. 3121*, May 1991.

[61]   E.L. Burgess and J.G. Fossum, "Performance of n$^+$p silicon solar cells in concentrated sunlight," *IEEE Trans. on Electron Devices*, vol. ED-24, pp. 433-438, Apr 1977.

[62]   System description information for 21 X concentrator supplied by Entech, Inc., Dallas, TX, Jan. 1994.

[63]   *Solar Radiation Energy Resource Atlas of the United States*, publication SERI/SP-642-1037, Solar Energy Research Institute (now NREL), Golden, CO, 1981.

[64]   A.B. Maish, "The status of photovoltaic concentrator development," *Proc. of the 11th E.C. Photovoltaic Solar Energy Conf.*, pp. 1644-1647, Oct. 1992.

[65]   P.J. Verlinden, et al., "A 26.8% efficient concentrator point-contact solar cell," *Conf. Proc. of the 13th European Photovoltaic Solar Energy Conf.*, pp. 1582-1585, Oct. 1995.

[66]   S. Khemthong, et al., "Fabrication experience with high efficiency silicon concentrator cells," *Conf. Rec. of the 13th IEEE Photovoltaic Specialists Conf.*, pp. 1046-1051, June 1978.

[67]   G. Stryi-Hipp, et al., "Precision spectral response and I-V characterization of concentrator cells," *Conf. Rec. of the 23rd IEEE Photovoltaic Specialists Conf.*, pp. 303-308, May 1993.

[68]   D.J. Friedman, et al., "GaInP/GaAs monolithic tandem concentrator cells," *Conf. Rec. of the First World Conf. on Photovoltaic Energy Conversion*, pp. 1829-1832, Dec. 1994.

[69]   J. Werth, "Design study of a thermophotovoltaic converter," *Proc. of the 18th Power Sources Conf.*, pp. 153-158, May 1964.

[70]   R.N. Bracewell and R.M. Swanson, "Silicon photovoltaic cells in TPV conversion," Report ER-633, Electric Power Research Institute, Feb. 1978.

[71]   W. Koechner, et al., *Technical and Economic Assessment of Solar Thermophotovoltaic Conversion*, pp. S6-S7, Electric Power Research Institute, July 1981.

[72]   L. Broman, "Thermophotovoltaics bibliography," *Progress in Photovoltaics*, vol. 3, pp. 65-74, Jan-Feb, 1995.

[73]   R. Siegel and J. R. Howell, *Thermal Radiation Heat Transfer*, Chap. 2, McGraw-Hill Book Co., New York, 1972.

[74]   R.C. Weast, ed., *Handbook of Chemistry and Physics*, 57th edition, CRC Press, Cleveland, p. E228, 1976.

[75]   B.L. Sater, "Verticle multi-junction cells for thermophotovoltaic conversion," *Conf. Rec. of the 1st NREL Conf. on Thermophotovoltaic Generation of Electricity*, AIP Conf. Proc. 321, pp. 165-176, July 1994.

[76]   L. Fraas, et al., "A thermophotovoltaic electric generator using GaSb cells with a hydrocarbon burner," *Conf. Rec. of the First World Conf. on Photovoltaic Energy Conversion*, pp. 1713-1716, Dec. 1994.

[77]   L.M. Fraas, et al., "Development of a small air-cooled 'Midnight Sun' thermophotovoltaic electric generator," *Conf. Rec. of the 2nd NREL Conf. on Thermophotovoltaic Generation of Electricity*, AIP Conf. Proc. 358, pp 128-133, July 1995.

[78]   D.L. Noreen and H. Du, "High power density thermophotovoltaic energy conversion," *Conf. Rec. of the 1st NREL Conf. on Thermophotovoltaic Generation of Electricity*, AIP Conf. Proc. 321, pp. 119-132, July 1994.

[79]   M.W. Wanless, et al., "$Ga_xIn_{1-x}As$ thermophotovoltiac converters," *Conf. Rec. of the First World Conf. on Photovoltiac Energy Conversion*, pp. 1685-1691, Dec. 1994.

[80]   R.E. Nelson and P.A. Iles, "Possible applications of selective emitters for space power," *Proc. of the 1993 ASME Joint Solar Engineering Conf.*, pp. 529-537, Apr. 1993.

[81]   G.E. Guazzoni, "High temperature spectral emittance of oxides of erbium, samarium, neodymium, and ytterbium," *Applied Spectroscopy*, vol. 26, pp. 60-65, 1972.

[82]   R.E. Nelson, "Grid-independent residential power systems," *Conf. Rec. of the 2nd NREL Conf. on Thermophotovoltaic Generation of Electricity*," AIP Conf. Proc. 358, pp. 221-237, July 1995.

[83]   W.J. Diederich, W. Newbury, and R.E. Nelson, U.S. Patent 4,883,619, "Refractory metal oxide processes," Nov. 28, 1989.

[84]   C.R. Parent, B.P. McFadden, and J.F. Olow, U.S. Patent 5,137,583, "Emission technology," Aug. 11, 1992.

[85]   Private communication with A. Kushch, May 1996.

INDEX